An Introduction to the
Mathematics of Money

David Lovelock   Marilou Mendel   A. Larry Wright

An Introduction to the Mathematics of Money

Saving and Investing

David Lovelock
Department of Mathematics
University of Arizona
Tucson, AZ 85721
USA
dsl@math.arizona.edu

Marilou Mendel
Department of Mathematics
University of Arizona
Tucson, AZ 85721
USA
mendel@math.arizona.edu

A. Larry Wright
Department of Mathematics
University of Arizona
Tucson, AZ 85721
USA
lwright@math.arizona.edu

Mathematics Subject Classification (2000): 91B82

ISBN 978-1-4419-2232-8   e-ISBN 978-0-387-68111-5
Printed on acid-free paper.

9 8 7 6 5 4 3 2 1

springer.com

Preface

Introduction

Some people distinguish between savings and investments, where savings are monies placed in relatively risk-free accounts with modest rewards, and where investments involve more risk and the potential for greater rewards. In this book we do not distinguish between these ideas. We treat them both under the umbrella of investing.

In general, income falls into two categories: EARNED INCOME—which is the income derived from your everyday job—and UNEARNED INCOME—which is income derived from investing. You attend college to strengthen your prospects for earned income, so why do you need to worry about unearned income, namely, investment income?

There are many reasons to invest and to learn about investing. Perhaps the primary one is to take charge of your own financial future. You need money for short-term goals (such as living expenses, emergencies) and for long-term goals (such as buying a car, buying a house, educating children, paying catastrophic medical bills, funding retirement).

Investing involves BORROWING AND LENDING, and BUYING AND SELLING.

- BORROWING AND LENDING. When you put money into a bank savings account, you are lending your money and the bank is borrowing it. You can lend money to a bank, a business, a government, or a person. In exchange for this, the borrower promises to pay you interest and to return your initial investment at a future date. Why would the borrower do this? Because the borrower anticipates using this money in a way that earns more than the interest promised to you. Examples of borrowing and lending are savings accounts, certificates of deposits, money-market accounts, and bonds.
- BUYING AND SELLING. When you buy something for investment purposes, you are buying an asset from a seller. You expect that this asset will generate a profit or will increase in value, part of which will be returned to you. Examples of this are owning real estate or stocks in companies.

There are two ways that you can make (lose) money buying stocks: stock-price appreciation (depreciation)—which depends on the expectations and opinions of the public—and dividends paid to you by the company—which depend on the company sharing its profits with you, a shareholder.

When investing, there are three things that can impact your profit—taxes, inflation, and risk. The first, taxes, should concern everyone. The second, inflation, should concern you if you make a profit. The third, risk, should concern you before you make an investment because risk influences the profitability of the investment. Generally, if you expect a high return on your money, then you should also expect a high risk. In the same way, low risks are usually associated with low returns. The larger the risk the greater the chance of actually losing money. There are various types of risk: inflation, market, currency fluctuations, political, interest-rate, liquidity, economic, default, business, etc.

Objectives and Background

We wrote this book with two objectives in mind:

- To use investing as a vehicle to introduce you, the student, to ideas, techniques, and applications that you might not encounter in your other mathematics courses. These include proofs by induction, recurrence relations, inequalities (in particular, the Arithmetic-Geometric Mean inequality and the Cauchy-Schwarz inequality), and elements of probability and statistics.
- To introduce you, the student, to elements of investing that are of life-long practical use. If you have not yet done so, then as you advance through life, you are forced to deal with such things as credit cards, student loans, car-loans, savings accounts, certificates of deposit, money-market accounts, mortgage payments, buying and selling bonds, and buying and selling stocks.

This book targets students at the sophomore/junior level, without assuming a background or any experience in investing. We assume knowledge of a two-semester calculus course as well as some mathematical sophistication. Specifically we use inequalities, log, exp, differentiation, the Mean Value Theorem, integration, Newton's method, limits of sequences, geometric series, the binomial expansion, and Taylor series.

There are problems at the end of each chapter. Some of these problems require that you have access to a spreadsheet program and that you know how to use it. A simple scientific or financial calculator (with functions such as log, exp, and the ability to calculate y^x) is all that is required for the remainder of the problems that involve arithmetical calculations. Some problems require you to obtain data from the World Wide Web (WWW), so access to the WWW, and familiarity with a browser, is a prerequisite.[1]

---

[1] The web page *www.mathematics-of-money.com* is dedicated to this book.

Comments

The following numbering system is used throughout the book: Example 2.3 refers to the third example in Chapter 2, Theorem 4.1 refers to the first theorem in Chapter 4, Figure 4.2 refers to the second figure in Chapter 4, Table 1.3 refers to the third table in Chapter 1, and Problem 1.5 refers to the fifth problem in Chapter 1.

The symbol $\triangle$ indicates the end of an example, and the symbol $\square$ indicates the end of a proof.

Many of the theorems in the book are given names (for example, The Compound Interest Theorem). This is done for ease of navigation for the student.

The problems are divided into two groups: "Walking", which involve routine, straight-forward calculations, and "Running", which are more challenging problems.

There are two appendices. Appendix A covers mathematical induction, recurrence relations, and inequalities. This material should be introduced at the beginning of the course. Appendix B covers elements of probability and statistics. It is not needed until the latter part of the course and can be introduced as needed. Many students may have seen this material in previous classes.

Unless indicated otherwise, all numerical results are rounded to three decimal places, and all dollar amounts are rounded to cents. Because of this convention, when the same calculation is performed in two different ways, the answers may differ slightly.

In most, but not all, cases in this book the interest rate is assumed to be positive. It is interesting to note that there are instances when the interest rate is negative. See, for example, [21]. A good reference on investments is [4]. A more advanced treatment is [18].

The information contained in this book is not intended to be construed as investment, legal, or accounting advice.

The Family

In order to try to personalize the investment examples and problems in this book, we have introduced a fictional family, the Kendricks. Helen (48) and Hugh (50) Kendrick, are husband and wife. They have three children, twins Wendy (25) and Tom (25), and Amanda (20), a college freshman. Jana Carmel (35) is one of Hugh's coworkers.

Acknowledgments

This book originated from classes taught in the Department of Mathematics at The University of Arizona, and in the Industrial Engineering and Operations Research Department at Columbia University. Several students provided valuable comments and corrected errors in the original notes. In particular, we thank Tom Wilkening and Michael Urbancic.

We also thank Wayne Hacker, David Lomen, and Doug Ulmer of the Department of Mathematics at The University of Arizona, who reviewed large portions of the manuscript and corrected several errors.

A guest lecture series, where professionals from both inside and outside academia are invited to discuss their specialities, is an integral part of the class and introduces the students to real-world applications of the mathematics of money. We thank the following guest lecturers: Dennis Bartlett, Sue Burroughs, Steve Kou, Steve Przewlocki, and Lauren Wright.

Special thanks go to Murray Teitelbaum and some of the many people at the New York Stock Exchange who are involved with the Teachers Workshop Program.

We thank the many other people who have assisted in the preparation of the manuscript: Pat Brockett, Han Gao, Joe Harwood, Pavan Korada, Robert Maier, Charles Newman, Keith Schlottman, Michael Sobel, and Aramian Wasielak.

We also thank the reviewers of the manuscript for their invaluable suggestions.

Finally, we thank our contacts at Springer—Achi Dosanjh, Yana Mermel, and Frank Ganz, for their indispensable support and advice.

Tucson, Arizona,   *David Lovelock*
July, 2006   *Marilou Mendel*
    A. Larry Wright

Contents

1

Simple Interest

Would you prefer to have $100 now or $100 a year from now? Even though the amounts are the same, most people would prefer to have $100 now because of the interest it can earn. Thus, whenever we talk of money we must state not only the amount, but also the time. This concept—that money today is worth more than the same amount of money in the future—is called the TIME VALUE OF MONEY. The PRESENT VALUE of an amount is its worth today, while the FUTURE VALUE is its worth at a later time. These topics are discussed here and in Chap. 2. Another reason that most people would prefer to have $100 now is that its purchasing power in the future may be less than at present due to inflation, which is discussed in Chap. 3.

When money earns interest it can do so in various ways—for example, simple interest, compounded annually, compounded semi-annually, compounded quarterly, compounded monthly, compounded daily, and compounded continuously. When referring to an interest rate, it is important to know which of these methods is being used.[1]

In this chapter we concentrate on simple interest. Compound interest is the subject of Chap. 2. A thorough familiarity with these two chapters is critical for an understanding of the rest of this book.

1.1 The Simple Interest Theorem

We invest $1,000 at 10% interest per year for 5 years. After one year we earn 10% of $1,000, namely, $100. We withdraw that interest and put it under a mattress, leaving the original $1,000 to earn interest in the second year. It too earns $100, which we also put under the mattress, so after two years we have the original $1,000 and $200 under the mattress. We continue doing this for 5 years, and so after five years we have the original $1,000 and $500 under

---

[1] A reference on interest rates with a historical summary dating back to about 400 B.C. is [15].

the mattress, for a total of $1,500. Table 1.1 shows the details. (Check the calculations in this table using a calculator or a spreadsheet program, and fill in the missing entries.)

Table 1.1. Simple Interest

	Year's Beginning	Year's End	
Year	Principal	Interest	Amount
1	$1,000.00	$100.00	$1,100.00
2	$1,000.00	$100.00	$1,200.00
3			
4	$1,000.00	$100.00	$1,400.00
5	$1,000.00	$100.00	$1,500.00

We now derive the general formula for this process. First, the total amount we have at any time is the **future value** of $1,000 **at that time**. Thus, $1,500 is the future value of $1,000 after 5 years. Second, the annual interest rate is called the **nominal rate**, the **quoted rate**, or the **stated rate**. Rather than restricting ourselves to annual calculations, we let n measure the total number of interest periods, of which we assume that there are m per year. (For example, if interest is calculated four times a year, that is, every three months, for five years, then $m = 4$ and $n = 4 \times 5 = 20$.) So we let[2]

P_0 be the INITIAL PRINCIPAL (present value, lump sum) invested,
n be the TOTAL NUMBER OF INTEREST PERIODS,
P_n be the FUTURE VALUE of P_0 at the end of the nth interest period,
m be the NUMBER OF INTEREST PERIODS PER YEAR,
$i^{(m)}$ be the NOMINAL RATE (annual interest rate), expressed as a decimal,
i be the INTEREST RATE PER INTEREST PERIOD.

The interest rate per interest period is

$$i = \frac{i^{(m)}}{m}.$$

For example, if the nominal rate is 12% calculated four times a year, then $m = 4$ and $i^{(4)} = 0.12$, so $i = 0.12/4 = 0.03$, the interest rate per quarter.

Using this notation we rewrite Table 1.1 symbolically in spreadsheet format[3] as Table 1.2, which is explained as follows.

---

[2] Throughout this chapter these symbols are used for this purpose. It is assumed that the units of currency are dollars, that m and n are positive integers, and that $i^{(m)} \geq 0$. Similar comments apply to subsequent chapters, as appropriate.

[3] We use this spreadsheet format throughout. It is always advisable to check calculations in more than one way—the spreadsheet is an excellent tool for this. The last entry on the Year 1 row, namely, $P_0 + iP_0 = P_1$, means that $P_0 + iP_0$ is the value of that entry, and we call it P_1.

Table 1.2. Simple Interest—Spreadsheet Format

Period	Period's Beginning Principal	Period's End Interest	Amount
1	P_0	iP_0	$P_0 + iP_0 = P_1$
2	P_0	iP_0	$P_1 + iP_0 = P_2$
3	P_0	iP_0	$P_2 + iP_0 = P_3$
4	P_0	iP_0	$P_3 + iP_0 = P_4$
5	P_0	iP_0	$P_4 + iP_0 = P_5$

At the end of the first interest period ($n = 1$) we receive iP_0 in interest, so the future value of P_0 after one period is

$$P_1 = P_0 + iP_0 = P_0 (1 + i).$$

At the end of the second interest period ($n = 2$) we again receive iP_0 in interest, so the future value after two periods is

$$P_2 = P_1 + iP_0 = P_0 (1 + i) + iP_0 = P_0 (1 + 2i).$$

At the end of the third interest period ($n = 3$) we again receive iP_0 in interest, so the future value after three periods is

$$P_3 = P_2 + iP_0 = P_0 (1 + 2i) + iP_0 = P_0 (1 + 3i).$$

This suggests the following theorem.

Theorem 1.1. *The Simple Interest Theorem.*
*If we start with principal P_0, and invest it for n interest periods at a nominal rate of $i^{(m)}$ (expressed as a decimal) calculated m times a year using **simple interest**, then P_n, the future value of P_0 at the end of n interest periods, is*

$$P_n = P_0 (1 + ni), \tag{1.1}$$

where $i = i^{(m)}/m$.

Proof. We can prove this theorem in at least two different ways: either using mathematical induction (see p. 245) or using recurrence relations (see p. 247).

We first prove it using mathematical induction. We know that (1.1) is true for $n = 1$. We assume that it is true for $n = k$, that is,

$$P_k = P_0(1 + ki),$$

and we must show that it is true for $n = k + 1$.

Now P_{k+1}, the amount of money at the end of period $k + 1$, is the sum of P_k, the amount of money at the beginning of this period, and iP_0, the interest earned during that period, that is,

$$P_{k+1} = P_k + iP_0,$$

so

$$P_{k+1} = P_0(1 + ki) + iP_0 = P_0(1 + (k+1)i),$$

which shows that (1.1) is true for $n = k + 1$. This concludes the proof by mathematical induction.

We now prove (1.1) using recurrence relations. We know that

$$P_{k+1} = P_k + iP_0,$$

so if we sum this from $k = 0$ to $k = n - 1$, then we have

$$\sum_{k=0}^{n-1} P_{k+1} = \sum_{k=0}^{n-1} P_k + \sum_{k=0}^{n-1} iP_0.$$

By canceling the common terms on both sides of this equation, we find that

$$P_n = P_0 + \sum_{k=0}^{n-1} iP_0 = P_0(1 + ni).$$

This concludes the proof using recurrence relations. $\square$

Comments About the Simple Interest Theorem

- We notice that $P_n = P_0(1 + ni)$ is a function of the three variables P_0, n, and i. We see that it is directly proportional[4] to P_0 and linear in each of the other two variables. Thus, a plot of the future value versus any one of these three variables, holding the other two fixed, is a line. An example of this is seen in Fig. 1.1, which shows the future value of \$1 as a function of n in years for 5% (the lower curve) and 10% (the upper curve) nominal interest rates $i^{(1)}$. You might ask why we selected $P_0 = 1$. We did this because P_n is directly proportional to P_0, so knowing the value of P_n when $P_0 = 1$ allows us to compute P_n for any other P_0, simply by multiplying by P_0. This is an important point, which recurs in later chapters.
- The quantity $P_n - P_0$ is the principal appreciation. Notice that, in the case of simple interest, $P_n - P_0 = P_0 ni$, that is, $P_n - P_0$ is directly proportional to P_0, n, and i, so doubling any of them doubles the principal appreciation. This is seen in Fig. 1.1. For example, if we look at $n = 20$, then we see that the vertical distance from the future value at 10% (\$3) to the present value (\$1) is twice the distance from the future value at 5% (\$2) to the present value (\$1).

---

[4] A function $f(x)$ is directly proportional to x if $f(x) = ax$, where a is a constant. "Directly proportional" is a special case of "linear".

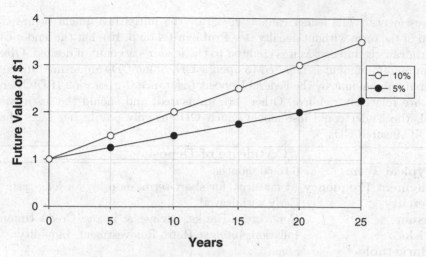

Fig. 1.1. Future Value of $1 at 5% and 10% simple interest

- The quantity $(P_n - P_0)/P_0$ is called the RATE OF RETURN. Notice that in the case of simple interest we have $(P_n - P_0)/P_0 = ni$, so doubling either n, the number of interest periods, or i, the interest rate per period, doubles the rate of return.

- The quantity $(P_n - P_0)/(nP_0)$ is the RATE OF RETURN PER PERIOD. Notice that in the case of simple interest we have $(P_n - P_0)/(nP_0) = i$, so the simple interest rate per interest period is the rate of return per period.

- Equation (1.1) is valid for $n \geq 0$. What happens if $n < 0$? In other words, if we accumulate P_0 over the past n years at a simple interest rate of i per year, then what amount, which we call P_{-n}, did we start with n years ago? From (1.1) we have

$$P_0 = P_{-n}(1 + ni),$$

so

$$P_{-n} = P_0 \frac{1}{1 + ni}, \tag{1.2}$$

which is not (1.1) with n replaced by $-n$. Thus, (1.1) is valid only for $n \geq 0$.

Financial Digression

A common form of investing is through a CERTIFICATE OF DEPOSIT (CD). CDs are issued by financial institutions. The institution pays a fixed interest rate on the lender's initial investment for a specified term. Typical terms are 6 months, and 1, 2, or 5 years. Usually the longer the term, the higher the rate because long-term investments are usually riskier than short-term investments.[5] The lender cannot withdraw the initial investment before the end of the term without penalty (see Problem 1.4 on p. 10), but the lender can withdraw the interest as it is credited to the lender's account, if desired. Often a minimum amount is required to open a CD. Some CDs are insured up to a maximum amount by the Federal Deposit Insurance Corporation (FDIC) and so are relatively risk-free. Others are uninsured, and should the institution fail, the lender could lose money. Such CDs usually pay higher rates than FDIC-insured CDs.

Certificate of Deposit	
Typical Term	6 to 60 months
Payment Frequency	At maturity for short-term; monthly for long-term
Penalty	Early withdrawal
Issuer	Commercial Banks, Savings & Loans, Credit Unions
Risks	Inflation, Interest Rate, Reinvestment, Liquidity
Marketable	Some
Restrictions	Minimum Investment

Example 1.1. Tom Kendrick invests $1,000 in a CD at 10% a year for five years. He withdraws the interest at the end of each year. What amount does he have at the end of five years assuming that he does not spend or invest the interest?

Solution. This is a simple interest example because the interest is withdrawn at the end of each year. Here the principal is $1,000 (so $P_0 = 1000$), the number of periods per year is 1 (so $m = 1$), the interest rate is 10% (so $i^{(1)} = 0.1$, and $i = i^{(1)}/m = 0.1$), and the number of years is 5 (so $n = 5$). Thus, the final amount is $P_5 = 1000(1 + 5(0.1)) = \$1,500$, which agrees with the step-by-step calculation on p. 2. $\triangle$

Example 1.2. Helen Kendrick invests $1,000 in a CD that doubles her money in five years. To what annual interest rate does this correspond assuming that she withdraws the interest each year?

Solution. Here the principal is $1,000 (so $P_0 = 1000$), the number of periods per year is 1 (so $m = 1$), the number of years is 5 (so $n = 5$), and the final amount is $2,000 (so $P_5 = 2000$). From (1.1) we have

$$2000 = 1000(1 + 5i),$$

so $i = 0.2$ and $i^{(1)} = mi = 0.2$, which is 20%.

We could also solve this using (1.2) with $P_0 = 2000$, $P_{-5} = 1000$, and $n = 5$, so that

$$1000 = 2000 \frac{1}{1 + 5i},$$

which again yields $i^{(1)} = mi = 0.2$. $\triangle$

---

[5] Examples of such risks are an institution defaulting on payment or an investor being locked in to a lower interest rate. Risks are discussed in greater detail in Chaps. 5 and 10.

Financial Digression

There are several investment vehicles available at banks and savings and loans in addition to CDs. The most common ones are savings accounts, checking accounts, and money market accounts.

SAVINGS ACCOUNTS pay a stated annual interest rate. In many cases, the interest is computed based on the daily balance. CHECKING ACCOUNTS may or may not pay interest. Both savings and checking accounts are "liquid", that is, the holder of the account may withdraw money at any time without penalty. Savings and checking accounts are frequently insured up to a maximum amount by the federal government.

The funds in a MONEY MARKET ACCOUNT are invested in vehicles such as short-term municipal bonds, Treasury bills,[6] and forms of short-term corporate debt. Money market accounts tend to pay a higher rate than savings accounts, checking accounts, or CDs. Money market accounts are not liquid in the sense that the number of withdrawals per month is limited.

The rates offered at different institutions for CDs, savings and checking accounts, and money market accounts are found in the financial sections of large city papers as well as financial newspapers such as the *Investor's Business Daily* and *The Wall Street Journal*.

Savings Account	
Typical Term	None
Payment Frequency	Monthly
Penalty	None
Issuer	Commercial Banks, Savings & Loans, Credit Unions
Risks	Reinvestment
Marketable	No
Restrictions	None

Example 1.3. Helen Kendrick has a savings account that pays interest at a nominal rate of 5%. Interest is calculated 365 times per year on the minimum daily balance and credited to the account at the end of the month. Helen has an opening balance of $1,000 at the beginning of April. On April 11 she deposits $200, and on April 21 she withdraws $300. How much interest does she earn in April?

Solution. Here $i^{(m)} = 0.05$ and $m = 365$, so $i = 0.05/365$. From April 1 to the end of April 11 Helen has $1,000 in the bank, so the interest earned is $1000(1 + 11(0.05/365)) - 1000 = \$1.51$.[7] However, she does not receive this $1.51 until the month's end. From April 12 to the end of April 20 Helen has $1,200 in the bank, so the interest earned is $1200(1 + 9(0.05/365)) - 1200 = \$1.48$. However, she does not receive this $1.48 until the month's end. From

---

[6] We discuss Treasury bills in Section 8.5.

[7] Note that interest is computed on the minimum daily balance. On April 11 the minimum balance is $1,000.

April 21 to the end of April 30 Helen has $900 in the bank, so the interest earned is $900\,(1 + 10(0.05/365)) - 900 = \$1.23$. At this stage Helen receives $\$1.51 + \$1.48 + \$1.23 = \$4.22$ in total interest.[8] $\triangle$

1.2 Ambiguities When Interest Period is Measured in Days

From a mathematical point of view, there is no ambiguity in calculating n, the total number of interest periods, and m, the number of interest periods per year. However, in practice, these quantities are ambiguous when the interest period is measured in days.

Number of Days Between Two Dates

There are different conventions used to calculate the total number of days between two dates. The most common are based either on the actual number of days between the dates or on a 30 day month.

The ACTUAL or EXACT NUMBER OF DAYS between two dates is calculated by counting the number of days between the given dates, excluding either the first or last day. Thus, the actual number of days between January 31 and February 5 is 5. Table 1.3 on p. 9 numbers the days of a year and is useful when computing the actual number of days.

Example 1.4. How many actual days are there between May 4, 2005 and October 3, 2005?

Solution. From Table 1.3, May 4 is day number 124 and October 3 is day number 276. So the actual number of days between them is $276 - 124 = 152$ days. $\triangle$

In the case of a leap year, there are two conventions: either February 29 is ignored, or it is included, in which case all numbers in Table 1.3 after February 28 are increased by one. In the actual method, February 29 is included.

The second convention, the 30-DAY MONTH METHOD, assumes that all months have 30 days. Here the number of days from the date $m_1/d_1/y_1$ to the date $m_2/d_2/y_2$, where m_i is the number of the month, d_i the day, and y_i the year of the date $(i = 1, 2)$, is given by the formula[9]

$$\text{Number of days} = 360\,(y_2 - y_1) + 30\,(m_2 - m_1) + (d_2 - d_1). \qquad (1.3)$$

---

[8] Because financial transactions are rounded to the nearest penny, all calculations are subject to roundoff error. It makes a difference whether the rounding is done before or after a calculation. For example, rounding $1.3698 + 1.6438 + 1.2328 = 4.2464$ after adding gives 4.25; rounding before gives $1.37 + 1.64 + 1.23 = 4.24$.

[9] Even this formula is not universally accepted. Sometimes additional conventions are adopted if either $d_1 = 31$ or $d_2 = 31$. (See Problem 1.9 on p. 11.)

Table 1.3. Numbered Days of the Year

Day	Jan	Feb	Mar	Apr	May	Jun	Jul	Aug	Sep	Oct	Nov	Dec
1	1	32	60	91	121	152	182	213	244	274	305	335
2	2	33	61	92	122	153	183	214	245	275	306	336
3	3	34	62	93	123	154	184	215	246	276	307	337
4	4	35	63	94	124	155	185	216	247	277	308	338
5	5	36	64	95	125	156	186	217	248	278	309	339
6	6	37	65	96	126	157	187	218	249	279	310	340
7	7	38	66	97	127	158	188	219	250	280	311	341
8	8	39	67	98	128	159	189	220	251	281	312	342
9	9	40	68	99	129	160	190	221	252	282	313	343
10	10	41	69	100	130	161	191	222	253	283	314	344
11	11	42	70	101	131	162	192	223	254	284	315	345
12	12	43	71	102	132	163	193	224	255	285	316	346
13	13	44	72	103	133	164	194	225	256	286	317	347
14	14	45	73	104	134	165	195	226	257	287	318	348
15	15	46	74	105	135	166	196	227	258	288	319	349
16	16	47	75	106	136	167	197	228	259	289	320	350
17	17	48	76	107	137	168	198	229	260	290	321	351
18	18	49	77	108	138	169	199	230	261	291	322	352
19	19	50	78	109	139	170	200	231	262	292	323	353
20	20	51	79	110	140	171	201	232	263	293	324	354
21	21	52	80	111	141	172	202	233	264	294	325	355
22	22	53	81	112	142	173	203	234	265	295	326	356
23	23	54	82	113	143	174	204	235	266	296	327	357
24	24	55	83	114	144	175	205	236	267	297	328	358
25	25	56	84	115	145	176	206	237	268	298	329	359
26	26	57	85	116	146	177	207	238	269	299	330	360
27	27	58	86	117	147	178	208	239	270	300	331	361
28	28	59	87	118	148	179	209	240	271	301	332	362
29	29		88	119	149	180	210	241	272	302	333	363
30	30		89	120	150	181	211	242	273	303	334	364
31	31		90		151		212	243		304		365

Example 1.5. How many days are there between May 4, 2005 and October 3, 2005 using the 30-day month convention?

Solution. Here $m_1 = 5$, $d_1 = 4$, $y_1 = 2005$, $m_2 = 10$, $d_2 = 3$, and $y_2 = 2005$, so (1.3) gives $360(2005 - 2005) + 30(10 - 5) + (3 - 4) = 149$ days. $\triangle$

Number of Days in a Year

There are also different conventions used to determine the number of days in the year. The two most common are the actual method (where the number of days is either 365 or 366) and the 30-day month method (where the number of days is computed from $12 \times 30 = 360$.)

When the actual method is used to calculate the number of days between two dates and the actual method is used to compute the number of days in a year, this is denoted by "actual/actual". Interest calculated using this convention is called **exact interest**.

When the 30-day month method is used to calculate the number of days between two dates and the 30-day month method is used to compute the number of days in a year, this is denoted by "30/360". Interest calculated using this convention is called **ordinary interest**.

When the actual method is used to calculate the number of days between two dates and the 30-day month method is used to compute the number of days in a year, this is denoted by "actual/360". Interest calculated using this convention is said to be computed by the **Banker's Rule**.

1.3 Problems

Walking

1.1. Tom Kendrick invests \$1,000 at a nominal rate of $i^{(1)}$, and he withdraws the interest at the end of each year. At the end of the fourth year he has earned \$300 in total interest. What nominal interest rate does he earn?

1.2. Tom Kendrick invests \$1,000 at a nominal rate of $i^{(2)}$, and he withdraws the interest at the end of each six months. At the end of the fourth year he has earned \$300 in total interest. What nominal interest rate does he earn? Would you expect it to be higher or lower than the answer to Problem 1.1?

1.3. Hugh Kendrick has a savings account that pays interest at a nominal rate of 3%. Interest is calculated 365 times a year on the minimum daily balance and credited to the account at the end of the month. Hugh has an opening balance of \$1,500 at the beginning of March. On March 13 he withdraws \$500, and on March 27 he deposits \$750. How much interest does he earn in March?

1.4. A certificate of deposit usually carries a penalty for early withdrawal: "The penalty is 90 days loss of interest, whether earned or not." Under what circumstances is it possible to lose money on a CD?

1.5. What is the actual number of days between October 4, 2004 and May 4, 2005?

1.6. What is the number of days between October 4, 2004 and May 4, 2005 using the 30-day month convention?

1.7. Explain why (1.3), namely $360\,(y_2 - y_1) + 30\,(m_2 - m_1) + (d_2 - d_1)$, gives the correct number of days between dates using the 30-day month convention.

1.8. Explain why the 30/360 method for calculating interest is unambiguous in a leap year.

1.9. A convention that is sometimes used to compute the number of days between two dates is based on 30-day month formula (1.3), namely $360(y_2 - y_1) + 30(m_2 - m_1) + (d_2 - d_1)$, but d_1 and d_2 are calculated from

$$d_i = \begin{cases} d_i \text{ if } 1 \le d_i \le 30, \\ 30 \text{ if } d_i = 31, \end{cases}$$

for $i = 1, 2$. This is sometimes referred to as the 30(E) method. Find two dates where the number of days between them differs using the 30-day month method and the 30(E) method.

Running

1.10. Show that simple interest calculated using exact interest is never greater than simple interest calculated using the Banker's Rule. Does a similar relationship hold between ordinary interest and the Banker's Rule? Explain.

Questions for Review

- What is meant by the expression "the time value of money"?
- What is the difference between the present value and the future value of money?
- How do you calculate simple interest?
- What is a proof by induction?
- What is a recurrence relation?
- Why is there ambiguity in counting the number of days between two dates?
- How do you count the number of days between two dates?
- What are the major differences between a CD, a savings account, a checking account, and a money market account?
- What is the rate of return on an investment?
- What does the Simple Interest Theorem say?

2

Compound Interest

The difference between simple interest and compound interest—the subject of this chapter—is that compound interest generates interest on interest, whereas simple interest does not.

2.1 The Compound Interest Theorem

We invest $1,000 at 10% per annum (per year), compounded annually for 5 years. After one year we earn 10% of $1,000 in interest, that is, $100. We combine that interest with the original amount, giving a new amount of $1,000 + $100 = $1,100. At the end of the second year this new amount earns 10% interest, that is, $110, giving a new amount of $1,100 + $110 = $1,210. If we continue doing this for 5 years, then at the end of the fifth year we have $1,610.51.[1] Table 2.1 shows the details. (Check the calculations in this table using a calculator or a spreadsheet program, and fill in the missing entries.)

Table 2.1. Compound Interest

Year	Year's Beginning Principal	Year's End Interest	Amount
1	$1,000.00	$100.00	$1,100.00
2	$1,100.00	$110.00	$1,210.00
3			
4	$1,331.00	$133.10	$1,464.10
5	$1,464.10	$146.41	$1,610.51

---

[1] Compare this with $1,000 invested for five years at 10% using simple interest. See Example 1.1 on p. 6.

We now derive the general formula for this process. As with simple interest, we let n measure the total number of interest periods, of which we assume that there are m per year. The total amount we have at the end of n interest periods is called the future value (or accumulated principal), and the annual interest rate is called the nominal rate. So we let[2]

P_0 be the INITIAL PRINCIPAL (present value, lump sum) invested,

n be the TOTAL NUMBER OF INTEREST PERIODS,

P_n be the FUTURE VALUE of P_0 (accumulated principal) at the end of the nth interest period,

m be the NUMBER OF INTEREST PERIODS PER YEAR,

$i^{(m)}$ be the NOMINAL RATE (annual interest rate), expressed as a decimal,

i be the INTEREST RATE PER INTEREST PERIOD.

The interest rate per interest period is $i = i^{(m)}/m$.

We want to find a formula for the future value P_n, and we do this by looking at $n = 1$, $n = 2$, and so on, hoping to see a pattern. Using this notation we rewrite Table 2.1 symbolically in spreadsheet format as Table 2.2, which is explained as follows.

Table 2.2. Compound Interest—Spreadsheet Format

Period	Period's Beginning Principal	Period's End Interest	Amount
1	P_0	iP_0	$P_0 + iP_0 = P_1$
2	P_1	iP_1	$P_1 + iP_1 = P_2$
3	P_2	iP_2	$P_2 + iP_2 = P_3$
4	P_3	iP_3	$P_3 + iP_3 = P_4$
5	P_4	iP_4	$P_4 + iP_4 = P_5$

At the end of the first interest period $(n = 1)$ we receive iP_0 in interest, so the future value of P_0 after one interest period is

$$P_1 = P_0 + iP_0 = P_0(1 + i).$$

At the end of the second interest period $(n = 2)$ we receive iP_1 in interest, so the future value of P_0 after two interest periods is

$$P_2 = P_1 + iP_1 = P_1(1 + i) = P_0(1 + i)^2.$$

At the end of the third interest period $(n = 3)$ we receive iP_2 in interest, so the future value of P_0 after three interest periods is

$$P_3 = P_2 + iP_2 = P_2(1 + i) = P_0(1 + i)^3.$$

---

[2] See footnote 2 on p. 2.

This suggests the following theorem.

Theorem 2.1. *The Compound Interest Theorem.*
If we start with principal P_0, and invest it for n interest periods at a nominal rate of $i^{(m)}$ (expressed as a decimal) compounded m times a year, then P_n, the future value of P_0 at the end of n interest periods, is

$$P_n = P_0(1+i)^n, \tag{2.1}$$

where $i = i^{(m)}/m$.

Proof. We can prove this theorem either by mathematical induction or by recurrence relations.

We first prove it using mathematical induction. We already know that (2.1) is true for $n = 1$. We assume that it is true for $n = k$, that is,

$$P_k = P_0(1+i)^k,$$

and we must show that it is true for $n = k + 1$.
  Now,

$$P_{k+1} = P_k + iP_k,$$

so

$$P_{k+1} = P_k(1+i) = P_0(1+i)^{k+1},$$

which shows that (2.1) is true for $n = k + 1$. This concludes the proof by mathematical induction.

We now prove (2.1) using recurrence relations. We know that

$$P_{k+1} = P_k + iP_k = (1+i)P_k,$$

so if we multiply this by $1/(1+i)^{k+1}$, then we can write it as

$$\frac{1}{(1+i)^{k+1}}P_{k+1} = \frac{1}{(1+i)^k}P_k.$$

Summing this from $k = 0$ to $k = n - 1$ gives

$$\sum_{k=0}^{n-1} \frac{1}{(1+i)^{k+1}}P_{k+1} = \sum_{k=0}^{n-1} \frac{1}{(1+i)^k}P_k,$$

or by canceling the common terms on both sides of this equation,

$$\frac{1}{(1+i)^n}P_n = P_0,$$

which is (2.1). This concludes the proof using recurrence relations. □

Comments About the Compound Interest Theorem

- We see that $P_n = P_0(1+i)^n$ is a function of the three variables P_0, n, and i. It is linear in P_0, but nonlinear in i and n. Thus, a plot of the future value versus either i or n, holding P_0 fixed, is not a line.
- Table 2.3 shows the future value of \$1 compounded annually (so $m = 1$) for different interest rates i and different numbers of years.

Table 2.3. Future Value of \$1

Interest Rate	Years					
	5	10	15	20	25	30
3%	1.159	1.344	1.558	1.806	2.094	2.427
4%	1.217	1.480	1.801	2.191	2.666	3.243
5%	1.276	1.629	2.079	2.653	3.386	4.322
6%	1.338	1.791	2.397	3.207	4.292	5.744
7%	1.403	1.967	2.759	3.870	5.427	7.612
8%	1.469	2.159	3.172	4.661	6.849	10.063
9%	1.539	2.367	3.643	5.604	8.623	13.268
10%	1.611	2.594	4.177	6.728	10.835	17.449
11%	1.685	2.839	4.785	8.062	13.586	22.892
12%	1.762	3.106	5.474	9.646	17.000	29.960
13%	1.842	3.395	6.254	11.523	21.231	39.115
14%	1.925	3.707	7.138	13.744	26.462	50.950
15%	2.011	4.046	8.137	16.367	32.919	66.212

- We can use Table 2.3 to show the dependence of P_n on i. This is seen in Fig. 2.1, which shows the future value of \$1 as a function of i for 10 years (the lower curve) and 20 years (the upper curve) with annual compounding. Both curves appear to be increasing and concave up. In Problem 2.26 on p. 40 you are asked to prove this.

- We can use Table 2.3 to show the dependence of P_n on n. This is seen in Fig. 2.2, which shows the future value of \$1 as a function of n for 5% interest (the lower curve) and 10% interest (the upper curve) compounded annually. Both curves appear to be increasing and concave up. In Problem 2.27 on p. 40 you are asked to prove this.

- Due to the linear relationship between $P_n - P_0$ and P_0, the principal appreciation, $P_n - P_0 = P_0 \left((1+i)^n - 1\right)$, doubles if P_0 doubles. What happens, however, when i or n doubles?
  - First, we discuss what happens to the principal appreciation if we double the interest rate, i. If we look at $n = 25$ in Fig. 2.2, then we see that the vertical distance from the future value at 10% (about \$10.80)

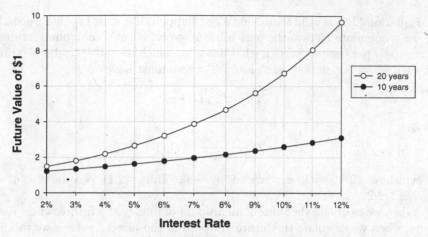

Fig. 2.1. Future Value of $1 for 10 and 20 years with annual compounding

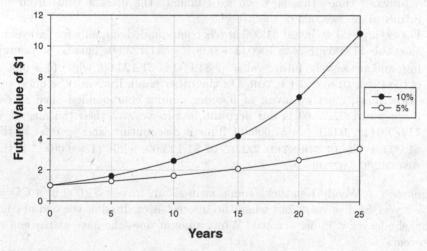

Fig. 2.2. Future Value of $1 at 5% and 10% interest compounded annually

to the present value ($1) is more than twice the distance from the future value at 5% (about $3.40) to the present value ($1). This suggests that doubling the interest rate more than doubles the principal appreciation. In Problem 2.28 on p. 40 you are asked to prove this.

○ Second, we discuss what happens to the principal appreciation if we double the number of periods n. If we look at the 10% curve in Fig. 2.2, then we see that the vertical distance from the future value at $n = 20$ (about $6.70) to the present value ($1) is more than twice the distance from the future value at $n = 10$ (about $2.60) to the present value ($1). This suggests that doubling the number of periods more than doubles the principal appreciation. In Problem 2.29 you are asked to prove this.

- Equation (2.1) is valid for $n \geq 0$. What happens if $n < 0$? In other words, if we accumulate P_0 over the past n interest periods at a compound interest rate of i per interest period, what amount, which we call P_{-n}, did we start with n interest periods ago? From (2.1) we must have

$$P_0 = P_{-n}(1 + i)^n,$$

so

$$P_{-n} = P_0 \frac{1}{(1 + i)^n} = P_0(1 + i)^{-n},$$

which is (2.1) with n replaced by $-n$. Thus, (2.1) is valid for $n = 0, \pm 1, \pm 2, \ldots$.

- When we calculate the value of an amount of money at a future time—that is, when we calculate the future value from the present value—we talk of COMPOUNDING. When we calculate the value of an amount of money at a previous time—that is, when we calculate the present value from the future value—we talk of DISCOUNTING.

  For example, if we invest $1,000 at 6% compounded annually for two years, then this $1,000 grows to $1000 (1 + 0.06)^2 = \$1,123.60$. This is compounding, and we say the future value of $1,000 is $1,123.60, while $(1 + 0.06)^2$ is the COMPOUNDING FACTOR. On the other hand, if we ask the question, "How much must we invest at 6% per annum compounded once a year if we want $1,123.60 in our account in two years?" then the answer is $1123.60 (1 + 0.06)^{-2} = \$1,000.00$. This is discounting, and we refer to the $1,000 as the DISCOUNTED VALUE of $1,123.60, while $(1 + 0.06)^{-2}$ is the DISCOUNT FACTOR.

Example 2.1. Wendy Kendrick, Tom's twin sister, invests $1,000 in a CD at 10% a year for five years, but when the interest is credited at the end of each year, she leaves it in her account. What amount does she have at the end of five years?

Solution. This is a compound interest example because the interest is not withdrawn, but earns interest. Here the principal is $1,000 (so $P_0 = 1000$), the compounding is once per year (so $m = 1$), the interest is 10% (so $i^{(1)} = 0.1$ and $i = i^{(1)}/1 = 0.1$), and the number of interest periods is 5 (so $n = 5$). Thus, the final amount is $P_5 = 1000(1 + 0.1)^5 = \$1,610.51$. This is $110.51 more than her brother made using simple interest in Example 1.1 on p. 6. $\triangle$

Example 2.2. Under the conditions of Example 2.1, find the future value if interest is compounded

(a) Semi-annually, that is, 2 times a year.
(b) Quarterly, that is, 4 times a year.
(c) Monthly, that is, 12 times a year.
(d) Daily, that is, 365 times a year.

Solution.

(a) In this case $i^{(2)} = 0.10$ and $m = 2$, so the semi-annual interest rate i is $i^{(2)}/2 = 0.10/2$. This is compounded 2×5 times, so the future value of P_0 after 5 years is

$$P_{10} = P_0 \left(1 + \frac{i^{(2)}}{2}\right)^{2\times5} = 1000 \left(1 + \frac{0.10}{2}\right)^{10} = \$1,628.89.$$

(b) In this case $i^{(4)} = 0.10$ and $m = 4$, so the quarterly interest rate i is $i^{(4)}/4 = 0.10/4$. This is compounded 4×5 times, so the future value of P_0 after 5 years is

$$P_{20} = P_0 \left(1 + \frac{i^{(4)}}{4}\right)^{4\times5} = 1000 \left(1 + \frac{0.10}{4}\right)^{20} = \$1,638.62.$$

(c) In this case $i^{(12)} = 0.10$ and $m = 12$, so the monthly interest rate i is $i^{(12)}/12 = 0.10/12$. This is compounded 12×5 times, so the future value of P_0 after 5 years is

$$P_{60} = P_0 \left(1 + \frac{i^{(12)}}{12}\right)^{12\times5} = 1000 \left(1 + \frac{0.10}{12}\right)^{60} = \$1,645.31.$$

(d) In this case $i^{(365)} = 0.10$ and $m = 365$, so the daily interest rate i is $i^{(365)}/365 = 0.10/365$. This is compounded 365×5 times, so the future value of P_0 after 5 years is

$$P_{1825} = P_0 \left(1 + \frac{i^{(365)}}{365}\right)^{365\times5} = 1000 \left(1 + \frac{0.10}{365}\right)^{1825} = \$1,648.61.$$

$\triangle$

This leads to the following result.
If P_0 is COMPOUNDED m TIMES A YEAR at a nominal interest rate of $i^{(m)}$, then the future value of P_0 after N years is

$$P_{mN} = P_0 \left(1 + \frac{i^{(m)}}{m}\right)^{mN}. \tag{2.2}$$

If we COMPOUND CONTINUOUSLY, by which we mean that the number of interest periods per year grows without bound, that is, $m \to \infty$, while the nominal rate $i^{(m)}$ is the same for all m, then, because $\lim_{m\to\infty} (1 + x/m)^m = e^u$ for all x,[0] the future value of P_0, denoted by P_∞, after N years at a nominal

---

[3] See Problem 2.31.

rate of $i^{(\infty)}$ is

$$P_\infty = P_0 e^{i^{(\infty)} N}.$$

So in Example 2.1 on p. 18, if \$1,000 is compounded continuously at 10% for 5 years, then we have $P_\infty = 1000 e^{0.1 \times 5} = \$1,648.72$.

If we tabulate these previous results with $i^{(m)} = 0.1$, then we have

m	Future Value
1	\$1,610.51
2	\$1,628.89
4	\$1,638.62
12	\$1,645.31
365	\$1,648.61
∞	\$1,648.72

From this table, and from our intuition, it appears that if the nominal rate $i^{(m)}$ is the same for all m, then the more frequently the compounding, the greater the future value. We justify this as follows.

Theorem 2.2. *If $i^{(m)}$ is positive and independent of m, $m \geq 1$, then the sequence $\left\{ \left(1 + \frac{i^{(m)}}{m} \right)^m \right\}$ is increasing, that is,*

$$\left(1 + \frac{i^{(m-1)}}{m-1} \right)^{m-1} < \left(1 + \frac{i^{(m)}}{m} \right)^m ,$$

and is bounded above by $e^{i^{(m)}}$, and $\lim_{m \to \infty} \left(1 + \frac{i^{(m)}}{m} \right)^m = e^{i^{(m)}}$.

Proof. To prove this we use the Arithmetic-Geometric Mean Inequality (see Appendix A.3 on p. 249), namely, if $a_1, a_2, \ldots, a_m$ are non-negative and not all zero, then

$$(a_1 a_2 \cdots a_m)^{1/m} \leq \frac{a_1 + a_2 + \cdots + a_m}{m},$$

with equality if and only if $a_1 = a_2 = \cdots = a_m$.

If we select $a_1 = 1$, $a_2 = a_3 = \cdots = a_m = 1 + i^{(m-1)}/(m-1)$, then we get[4]

$$\left(1 \left(1 + \frac{i^{(m-1)}}{m-1} \right)^{m-1} \right)^{1/m} < \frac{1 + (m-1)\left(1 + \frac{i^{(m-1)}}{m-1} \right)}{m},$$

or

$$\left(\left(1 + \frac{i^{(m-1)}}{m-1} \right)^{m-1} \right)^{1/m} < \frac{m + i^{(m-1)}}{m} = 1 + \frac{i^{(m-1)}}{m}.$$

---

[4] Equality cannot occur because $a_1 \neq a_2$.

But $i^{(m-1)} = i^{(m)}$ (why?), so

$$\left(1 + \frac{i^{(m-1)}}{m-1}\right)^{m-1} < \left(1 + \frac{i^{(m)}}{m}\right)^m.$$

Thus, the sequence $\left\{\left(1 + \frac{i^{(m)}}{m}\right)^m\right\}$ is increasing.

Because $\lim_{m\to\infty} \left(1 + i^{(m)}/m\right)^m = e^{i^{(m)}}$, the sequence is bounded above by $e^{i^{(m)}}$. $\quad\square$

In Problem 2.42 on p. 42, you are asked to prove a more general result than Theorem 2.2.

In order to compare investments with a constant rate of return but with different frequencies of compounding, it is common to calculate the Annual Effective Rate (EFF) for each investment.[5]

Definition 2.1. *The* ANNUAL EFFECTIVE RATE *(EFF),* i_{eff}, *is the annual rate of return* $i^{(1)}$ *that is equivalent to the nominal rate* $i^{(m)}$ *(compounded m times a year), or the nominal rate* $i^{(\infty)}$ *(compounded continuously).*[6]

If the investment is compounded m times a year, then this means that

$$P_0(1 + i_{\text{eff}}) = P_0\left(1 + \frac{i^{(m)}}{m}\right)^m,$$

so

$$i_{\text{eff}} = \left(1 + \frac{i^{(m)}}{m}\right)^m - 1 = (1+i)^m - 1. \tag{2.3}$$

If the investment is compounded continuously, then this means that

$$P_0(1 + i_{\text{eff}}) = P_0 e^{i^{(\infty)}},$$

so

$$i_{\text{eff}} = e^{i^{(\infty)}} - 1. \tag{2.4}$$

Example 2.3. Washington Federal Savings offers a CD with a nominal rate of 4.88% compounded 365 times a year. What is the EFF?

---

[5] In Sect. 2.3, we discuss how to compare investments that are not annual. This comparison requires introducing the Internal Rate of Return of an investment.

[6] The annual effective rate is sometimes called the Annual Percentage Rate (APR) when one is referring to debts. On financial calculators, the Annual Effective Rate is often calculated using the EFF button.

Solution. From (2.3), with $i^{(365)} = 0.0488$ and $m = 365$, we have

$$i_{\text{eff}} = \left(1 + \frac{0.0488}{365}\right)^{365} - 1 = 0.05.$$

So $i_{\text{eff}} = 5\%$. $\triangle$

Example 2.4. Wendy Kendrick has the choice between two CDs, both of which mature in one year. One offers a nominal rate of 8% compounded semi-annually, and the other 7.85% compounded 365 times a year. Which is the better deal?

Solution. With $i^{(2)} = 0.08$, we have $i = 0.08/2$, so

$$i_{\text{eff}} = \left(1 + \frac{0.08}{2}\right)^2 - 1 = 0.0816.$$

With $i^{(365)} = 0.0785$, we have $i = 0.0785/365$, so

$$i_{\text{eff}} = \left(1 + \frac{0.0785}{365}\right)^{365} - 1 = 0.0817.$$

The second CD is the better deal.

An alternative way of answering this question is to compute the future value of each CD assuming an initial investment of P_0. The future value of P_0 in the first case is $P_0 \left(1 + 0.08/2\right)^2 = 1.0816 P_0$, while in the second case it is $P_0 \left(1 + 0.0785/365\right)^{365} = 1.0817 P_0$. $\triangle$

Example 2.5. Henry Kendrick's business can buy a piece of equipment for $200,000 now, or for $70,000 now, $70,000 in one year, and $70,000 in two years. Which option is better if money can be invested at a nominal rate of 6% compounded monthly?

Solution. We can solve this in two different ways, which shed light on the concept of present value.

Solution 1.

The present value of the three cash flows is

$$70000 + 70000 \left(1 + \frac{0.06}{12}\right)^{-12} + 70000 \left(1 + \frac{0.06}{12}\right)^{-24} = \$198{,}036.37.$$

Thus, the present value is less than $200,000 so he would save $200,000 − $198,036.37, that is, $1,963.63, which has a future value of $1963.63(1 + 0.06/12)^{24} = \$2{,}213.32$. Thus, the installment plan is better.

Solution 2.

In order to consider these two options, Henry's business must have $200,000 available. So under the installment plan, he first pays $70,000, leaving $130,000. He invests this at 6% for one year, giving $138,018.12. He then pays $70,000, leaving $68,018.12. This is invested for one year, giving $72,213.33. After paying $70,000 he is left with $2,213.33, which has a present value of $1,963.63.

If we let $P = 200000$, $M = 70000$, and $i = 0.06/12$, then we can see the equivalence of these two approaches because

$$P - M - M(1+i)^{-12} - M(1+i)^{-24}$$
$$= \left(\left(\left((P-M)(1+i)^{12} - M \right)(1+i)^{12} - M \right)(1+i)^{-24} \right.$$

$\triangle$

2.2 Time Diagrams and Cash Flows

A useful device, called a TIME DIAGRAM, allows us to visualize the CASH FLOW—the flow of cash in and out of an investment.

- First, a horizontal line is drawn, which represents time increasing from the present (denoted by 0) as we move from left to right.
- Second, we draw short vertical lines that start on the horizontal line. Those that go up represent cash coming in (a positive cash flow, or receipts), while those that go down represent cash going out (a negative cash flow, or disbursement). Thus, the vertical lines represent the cash flow.

For example, suppose that we invest $1,000 at 6% compounded annually for two years. At the end of two years this $1,000 grows to the future value $1000(1 + 0.06)^2 = \$1,123.60$. The cash flows are represented as follows:

Years	0	1	2
Cash Flow	−$1,000.00	$0.00	+$1,123.60

At year zero we invested $1,000 (so the cash went out, and hence the minus sign), and at year two we received $1,123.60 (so the cash came in, and hence the plus sign, which we normally omit). We represent this with the time diagram shown in Fig. 2.3.

In general, the cash flows for compounding are

Time	0	1	$\cdots$	$n-1$	n
Cash Flow	$-P$	0	$\cdots$	0	$P(1+i)^n$

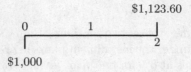

Fig. 2.3. Time diagram

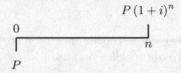

Fig. 2.4. Time diagram for compounding

and are represented by Fig. 2.4.

The cash flows for discounting are

Time	0	1	$\cdots$	$n-1$	n
Cash Flow	$-P(1+i)^{-n}$	0	$\cdots$	0	P

and are represented by Fig. 2.5.

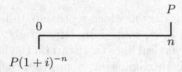

Fig. 2.5. Time diagram for discounting

If we consider Fig. 2.6, then we see that it represents the following cash

Fig. 2.6. Mystery time diagram

flows: initially an amount P is invested, and at unit time intervals an amount F is received regularly, and finally, after n time intervals, an amount $F + P$ is received, that is,

Time	0	1	2	$\cdots$	$n-1$	n
Cash Flow	$-P$	F	F	$\cdots$	F	$F+P$

The following net cash flows represent the general case,

Time	0	1	2	$\cdots$	$n-1$	n
Cash Flow	C_0	C_1	C_2	$\cdots$	C_{n-1}	C_n

where C_k $(k = 0, 1, \ldots, n)$ are positive, negative, or zero. In this case, without knowing the sign of C_k we cannot, on the time diagram, correctly indicate whether C_k should be above or below the time line. We use Fig. 2.7 to represent this general case.

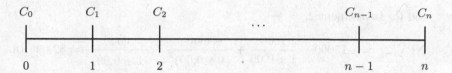

Fig. 2.7. Time diagram for general cash flows

The NET PRESENT VALUE, NPV, of an investment is the difference between the present value of the cash inflows and the present value of the cash outflows, that is,

$$\text{NPV} = C_0 + C_1(1+i)^{-1} + C_2(1+i)^{-2} + \cdots + C_n(1+i)^{-n},$$

where i is the prevailing interest rate. (Usually the initial cash flow, C_0, is negative.) This interest rate is a function of the risk of the investment. When attempting to choose between two investments with the same risks, the investor generally chooses the one with the higher net present value. If both investments have the same net present value and the same time interval, then the investor is said to be INDIFFERENT BETWEEN the investments.

Example 2.6. Tom Kendrick is considering two investments with annual cash flows

Years	0	1	2	3
Cash Flow (Investment 1)	−$13,000	$5,000	$6,000	$7,000
Cash Flow (Investment 2)	−$13,000	$7,000	$4,800	$6,000

Which is the better investment if the prevailing interest rate is (a) 4.5%? (b) 9%?

Solution.

(a) At 4.5%, the NPV for Investment 1 is

$$\text{NPV}(1) = -13000 + \frac{5000}{1+0.045} + \frac{6000}{(1+0.045)^2} + \frac{7000}{(1+0.045)^3} = \$3{,}413.14,$$

and for Investment 2,

$$NPV(2) = -13000 + \frac{7000}{1+0.045} + \frac{4800}{(1+0.045)^2} + \frac{6000}{(1+0.045)^3} = \$3{,}351.85.$$

Thus, at 4.5%, Investment 1 is the better choice.

(b) At 9%, the NPV for Investment 1 is

$$NPV(1) = -13000 + \frac{5000}{1+0.09} + \frac{6000}{(1+0.09)^2} + \frac{7000}{(1+0.09)^3} = \$2{,}042.52,$$

and for Investment 2,

$$NPV(2) = -13000 + \frac{7000}{1+0.09} + \frac{4800}{(1+0.09)^2} + \frac{6000}{(1+0.09)^3} = \$2{,}095.18.$$

Thus, at 9%, Investment 2 is the better choice.

$\triangle$

In this example we see that increasing the prevailing interest rate causes the NPV of a cash flow to drop. This is generally true if $C_0 < 0$ and $C_1, C_2, \ldots, C_n$ are non-negative, but not all zero.

2.3 Internal Rate of Return

If we invest $1,000 at 6% per annum, and then a year later invest $2,000 at 5% per annum, what is the future value of the entire investment after a total of two years? This is represented by the following cash flows,

Years	0	1	2
Cash Flow	−$1,000.00	−$2,000.00	?

or by Fig. 2.8.

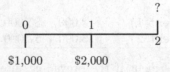

Fig. 2.8. Time diagram

The $1,000 has a future value of $1000(1+0.06)^2 = \$1{,}123.60$ after 2 years, while the $2,000 has a future value of $2000(1+0.05) = \$2{,}100$ after 1 year, so the future value of the entire investment is

$$1000(1+0.06)^2 + 2000(1+0.05) = 1123.60 + 2{,}100 = \$3{,}223.60.$$

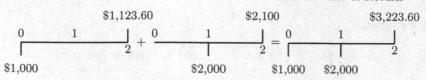

Fig. 2.9. Decomposition of time diagram

Figure 2.9 shows how to decompose Fig. 2.8.

Because the interest rates changed during the investment period, a natural question to ask is, "What rate have we really been earning over the two years?" (We cannot use the annual effective rate i_{eff} because that applies only to investments that earn a constant rate of return.) One way to answer this question is to say that we made \$223.60 on an investment of \$3,000 over two years, so we made \$223.60/\$3,000 = 0.0745 over two years, which is a rate of 0.03725 per year.

However, there are two things wrong with this. First, by dividing 0.0745 by 2 we have computed a simple interest rate. Second, we have not taken into account that the \$2,000 and the \$1,000 are deposited at different times.

We can correct the first problem by finding the annual interest rate i required to discount \$3,223.60 to \$3,000, that is, find $i > 0$ for which

$$3000 = 3223.60 \left(1 + i\right)^{-2},$$

so

$$i = \pm\sqrt{\frac{3223.60}{3000}} - 1 = 0.0366 \text{ or } -2.0366.$$

Because i must be positive, we find that $i = 0.0366$. (Why is this way of computing the interest rate lower than 0.03725, the interest rate we obtained the previous way?)

However, this technique does not take into account the second problem, namely, that the \$2,000 was deposited at a different time from the \$1,000. We can solve this problem by finding the annual interest rate r required to discount \$3,223.60 to \$1,000 plus the discounted value of \$2,000, namely $2000(1 + r)^{-1}$. Thus, we must find r for which

$$1000 + 2000(1 + r)^{-1} = 3223.60(1 + r)^{-2}.$$

This means we must solve the quadratic equation

$$1000(1 + r)^2 + 2000(1 + r) - 3223.60 = 0,$$

that is,

$$(1 + r)^2 + 2(1 + r) - 3.2236 = 0,$$

for $1 + r$, giving

$$1 + r = \frac{1}{2}\left(-2 \pm \sqrt{2^2 + 4 \times 3.22360}\right) = 1.055 \text{ or } -3.055,$$

so $r = 0.055$ or -4.055. However, r must be positive, so $r = 0.055$. The annual interest rate computed in this way is called the INTERNAL RATE OF RETURN (IRR), and takes into account both compounding and the time value of money.

Definition 2.2. *The* INTERNAL RATE OF RETURN *(IRR)*, i_{irr}, *for an investment is the interest rate that is equal to the annually compounded rate earned on a savings account with the same cash flows.*

Example 2.7. Hugh Kendrick, the father of Wendy and Tom, invests $10,000 in a CD that yields 5% compounded 365 days a year. At the end of the year he moves the proceeds to a new CD that yields 6% compounded 4 times a year. How much interest does he have when the second CD matures? What is the IRR, i_{irr}, for this investment?

Solution. At the end of the first year he has $10000\left(1 + 0.05/365\right)^{365}$, which is the amount he invests as the principal in the second CD. This leads to

$$\left(10000\left(1 + \frac{0.05}{365}\right)^{365}\right)\left(1 + \frac{0.06}{4}\right)^4 = \$11{,}157.77,$$

so he earned $1,157.77 in interest.

To find the IRR, we want to find the rate i_{irr} for which

$$10000(1 + i_{irr})^2 = 10000\left(1 + \frac{0.05}{365}\right)^{365}\left(1 + \frac{0.06}{4}\right)^4,$$

so

$$i_{irr} = \pm\sqrt{\left(1 + \frac{0.05}{365}\right)^{365}\left(1 + \frac{0.06}{4}\right)^4} - 1 = 0.0563 \text{ or } -2.0563.$$

Thus, the IRR is about 5.6%. △

Example 2.8. Find the IRR, i_{irr}, that is equivalent to a simple interest investment rate of 20% a year for 5 years.

Solution. Over 5 years under simple interest with $P_0 = 1$, $i = 0.20$, and $n = 5$, (1.1) gives

$$P_5 = (1 + (5 \times 0.2)) = 2.$$

Now, $(1 + i_{irr})^5 = 2$ means that $i_{irr} = 2^{1/5} - 1 = 0.149$.

Thus, a rate of 14.9%, compounded annually, doubles an investment in 5 years. △

Financial Digression
An INDEX FUND is a fund whose collection of assets is designed to mirror the performance of a particular broad-based stock market index.[7] Usually index funds are managed so that decisions are automated and transactions are infrequent. There is usually a minimum opening balance, unless the investor invests a regular amount each month. Investors are usually discouraged from frequent buying and selling, and they may be penalized for this. The first index fund for individual investors was created by The Vanguard Group in 1976.

Example 2.9. At the beginning of every month for 12 months, Hugh Kendrick buys $100 worth of shares in an index fund. At the end of the twelfth month his shares are worth $1,500. What is the internal rate of return, i_{irr}, of his investment?

Solution. Hugh's investment is represented by Fig. 2.10.

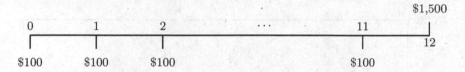

Fig. 2.10. Internal rate of return

The present value of his investment at an annual interest rate of i_{irr} is

$$100 + 100(1 + i_{\text{irr}})^{-1/12} + 100(1 + i_{\text{irr}})^{-2/12} + \cdots + 100(1 + i_{\text{irr}})^{-11/12}$$
$$= 1500(1 + i_{\text{irr}})^{-12/12},$$

or

$$100(1 + i_{\text{irr}})^{12/12} + 100(1 + i_{\text{irr}})^{11/12} + \cdots + 100(1 + i_{\text{irr}})^{1/12} = 1500.$$

The first equation is obtained by discounting to the present value, the second by compounding to the future value at the end of the twelfth month. Notice we assume that $1 + i_{\text{irr}} > 0$.

If in the second equation we let $1 + i = (1 + i_{\text{irr}})^{1/12}$, so $1 + i > 0$, then we find that

$$100(1 + i)^{12} + 100(1 + i)^{11} + 100(1 + i)^{10} + \cdots + 100(1 + i) = 1500. \quad (2.5)$$

We rewrite the equation as

$$\left((1 + i)^{11} + (1 + i)^{10} + \cdots + 1\right)(1 + i) = 15,$$

---

[7] We discuss stock market indexes in Chap. 10.

and then use the geometric series[8]

$$1 + x + x^2 + \cdots + x^{n-1} = (x^n - 1)/(x - 1), \text{ valid for } x \neq 1 \text{ and } n \geq 1,$$

with $x = 1 + i$ and $n = 12$, to find that i satisfies

$$\frac{(1+i)^{12} - 1}{i}(1+i) - 15 = 0. \tag{2.6}$$

In general we are unable to solve the equation for i analytically, but it can be solved numerically—for example, by the bisection method or by Newton's Method (see Problem 2.20 on p. 39)—or graphically, by graphing the left-hand side of (2.6) and estimating where it crosses the horizontal axis (see Fig. 2.11), giving $i = 0.0339$. Here $1 + i = 1.0339 > 0$ and from $1 + i = (1 + i_{irr})^{1/12}$ we find that $i_{irr} = (1 + i)^{12} - 1 = 0.492$. Thus, the internal rate of return is 49.2%. △

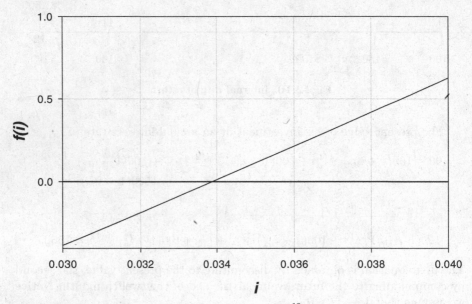

Fig. 2.11. The function $f(i) = \frac{(1+i)^{12}-1}{i}(1+i) - 15$

In the previous example we have overlooked a very important point—how do we know that $1 + i = 1.0339$ is the only solution of (2.6)? We don't, and it isn't, because $1 + i = -1.318$ is another solution. But this solution does not satisfy $1 + i > 0$, so we reject it. So how do we know that 1.0339 is the only acceptable solution?

---

[8] See Problems 2.34 and 2.35.

In fact, if we rewrite (2.5) in the form

$$100(1+i)^{12} + 100(1+i)^{11} + 100(1+i)^{10} + \cdots + 100(1+i) - 1500 = 0, \quad (2.7)$$

then we see that this is a polynomial equation of degree 12 in $(1+i)$. In general, polynomials of degree n have n roots (some of which may be complex); thus, we might expect (2.7) to have 12 solutions! In fact, it has only one real solution that satisfies $1 + i > 0$, which we show shortly.

To show this, we turn to the general case, where we have the following net cash flows:

Period	0	1	2	$\cdots$	$n-1$	n
Cash Flow	C_0	C_1	C_2	$\cdots$	C_{n-1}	C_n

where C_k $(k = 0, 1, \ldots, n)$ are positive, negative, or zero. We let m be the number of periods per year, while n is the total number of periods. Figure 2.12 represents this general case.

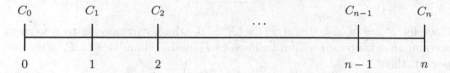

Fig. 2.12. General time diagram

The IRR, i_{irr}, for this series of cash flows is the solution of the equation[9]

$$C_0 + C_1(1 + i_{\text{irr}})^{-1/m} + C_2(1 + i_{\text{irr}})^{-2/m} + \cdots + C_n(1 + i_{\text{irr}})^{-n/m} = 0,$$

which can be written in the form

$$C_0(1 + i_{\text{irr}})^{n/m} + C_1(1 + i_{\text{irr}})^{(n-1)/m} + C_2(1 + i_{\text{irr}})^{(n-2)/m} + \cdots + C_n = 0.$$

If we let $1 + i = (1 + i_{\text{irr}})^{1/m}$, where $1 + i > 0$, then this last equation becomes

$$C_0(1 + i)^n + C_1(1 + i)^{n-1} + C_2(1 + i)^{n-2} + \cdots + C_n = 0.$$

This is a polynomial equation of degree n in $1 + i$, and therefore has exactly n solutions, including complex ones.

---

[9] Notice that the IRR for an investment is the annual rate that makes the net present value equal to zero. So the IRR is the rate that makes the present value of the expected future cash flows equal to the initial cost of the investment. The IRR is useful for comparing investments with different costs or with cash flows that differ in terms of amount or frequency of payment, and for determining how a potential investment compares with the investor's requirements.

From $1 + i = (1 + i_{\text{irr}})^{1/m}$ we find that

$$i_{\text{irr}} = (1 + i)^m - 1,$$

which relates the internal rate of return, i_{irr}, to the periodic internal rate of return, i. Notice that if $m = 1$, then $i_{\text{irr}} = i$, and that if $i > 0$, then $i_{\text{irr}} > 0$. If we use the binomial approximation

$$(1 + i)^m = 1 + mi + \cdots,$$

then we find that

$$i_{\text{irr}} \approx mi,$$

which is sometimes used in place of $i_{\text{irr}} = (1 + i)^m - 1$.

For the case $n = 1$ (so there are two cash flows), this equation reduces to

$$C_0(1 + i) + C_1 = 0,$$

which has the solution

$$i = \frac{-C_1 - C_0}{C_0}.$$

If we let $P_0 = -C_0 < 0$ and $P_1 = C_1 > 0$, which corresponds to a savings account that is opened with a balance of P_0 and accumulates to P_1 after one period, then

$$i = \frac{P_1 - P_0}{P_0}.$$

This is the rate of return defined on p. 5 for these cash flows. Thus, **if there is one cash inflow and one cash outflow, then the periodic internal rate of return and the rate of return of the investment are identical.**

For the case $n = 2$ this equation reduces to

$$C_0 i^2 + (2C_0 + C_1) i + (C_0 + C_1 + C_2) = 0,$$

which could have no real solutions, one real solution repeated, or two distinct real solutions. Only in the second and third cases does i exist and is possibly unique. In fact, it is possible to construct perfectly reasonable cash flows for which the IRR does not exist, or does exist, but is not unique.

Example 2.10. Find the IRR for

Years	0	1	2
Cash Flow	−$1,000	$2,000	−$1,500

This corresponds to giving someone $1,000 now and $1,500 after two years, in exchange for $2,000 after one year.

Solution. Here i_{irr} must satisfy

$$-1000(1 + i_{\text{irr}})^2 + 2000(1 + i_{\text{irr}}) - 1500 = 0,$$

which reduces to

$$2i_{\text{irr}}^2 + 1 = 0,$$

which has no real solutions. Thus, this transaction has no IRR. $\triangle$

Example 2.11. Find the IRR for

Years	0	1	2
Cash Flow	−$1,000	$2,150	−$1,155

This corresponds to giving someone $1,000 now and $1,155 after two years, in exchange for $2,150 after one year.

Solution. Here i_{irr} must satisfy

$$-1000(1 + i_{\text{irr}})^2 + 2150(1 + i_{\text{irr}}) - 1155 = 0,$$

which, when solved for $1 + i_{\text{irr}}$, reduces to

$$1 + i_{\text{irr}} = 1.05 \text{ or } 1.1.$$

Thus, $i_{\text{irr}} = 0.05$ or $i_{\text{irr}} = 0.10$, and i_{irr} is not unique. $\triangle$

We can give partial answers to the uniqueness question as follows.

Theorem 2.3. *The IRR Uniqueness Theorem I.*
If there exists an integer p for which $C_0, C_1, \ldots, C_p$ are of the same sign or zero (but not all zero) and for which $C_{p+1}, C_{p+2}, \ldots, C_n$ are all of the opposite sign or zero (but not all zero), then

$$C_0 + C_1(1 + i)^{-1} + C_2(1 + i)^{-2} + \cdots + C_n(1 + i)^{-n} = 0$$

has at most one positive solution $1 + i$.

Proof. By thinking of

$$C_0(1 + i)^n + C_1(1 + i)^{n-1} + C_2(1 + i)^{n-2} + \cdots + C_n = 0$$

as a polynomial equation in $1 + i$, and using Descartes' Rule of Signs,[10] there is exactly one change of sign, so there is at most one positive root of the polynomial. $\square$

---

[10] Descartes' Rule of Signs states that the maximum number of positive solutions of the polynomial equation $a_n x^n + a_{n-1} x^{n-1} + \cdots + a_1 x + a_0 = 0$ is the number of sign changes that occur when one looks at the coefficients $a_n, a_{n-1}, \ldots, a_1, a_0$ in order (excluding the coefficients that are zero).

Comments About the IRR Uniqueness Theorem I

- This theorem guarantees that there is at most one positive solution, $1+i$. However, that does not guarantee that i is positive.
- Returning to (2.7) we see that it has one change of sign, so it has at most one positive solution—the one we found numerically. Each of the two examples for which the IRR either did not exist (Example 2.10 on p. 32) or was not unique (Example 2.11 on p. 33) had two sign changes.

There is another theorem that is sometimes useful (see [16]).

Theorem 2.4. *The IRR Uniqueness Theorem II.*
If there exists an i for which

(a) $1 + i > 0$,
(b) $\sum_{k=0}^{p} C_k(1+i)^{p-k} > 0$ *for all integers p satisfying $0 \le p \le n-1$ (that is, the future value of all the cash flows up to period p are positive), and*
(c) $\sum_{k=0}^{n} C_k(1+i)^{n-k} = 0$,

then i is unique.

Proof. Assume that there is a second solution j of (c), that is,

$$\sum_{k=0}^{n} C_k(1+j)^{n-k} = 0,$$

satisfying (a) and (b). Without loss of generality, we may assume that $j > i$.

We first prove, by induction on p, that

$$\sum_{k=0}^{p} C_k(1+j)^{p-k} > \sum_{k=0}^{p} C_k(1+i)^{p-k}$$

for $p = 1$ to n. For $p = 1$ this becomes

$$C_0(1+j) + C_1 > C_0(1+i) + C_1,$$

which is true because from condition (b) with $p = 0$, we have $C_0 > 0$.

We assume that $\sum_{k=0}^{h} C_k(1+j)^{h-k} > \sum_{k=0}^{h} C_k(1+i)^{h-k}$, and we must show that

$$\sum_{k=0}^{h+1} C_k(1+j)^{h+1-k} > \sum_{k=0}^{h+1} C_k(1+i)^{h+1-k}.$$

Now,

$$\sum_{k=0}^{h+1} C_k(1+j)^{h+1-k} = \sum_{k=0}^{h} C_k(1+j)^{h+1-k} + C_{h+1}$$

$$= \sum_{k=0}^{h} C_k(1+j)^{h-k}(1+j) + C_{h+1}$$

$$> \sum_{k=0}^{h} C_k(1+i)^{h-k}(1+j) + C_{h+1}$$

$$> \sum_{k=0}^{h} C_k(1+i)^{h-k}(1+i) + C_{h+1}$$

$$= \sum_{k=0}^{h+1} C_k(1+i)^{h+1-k},$$

which proves the inequality by induction. (Where did we use condition (a)? Condition (b)?)

Thus,

$$\sum_{k=0}^{n} C_k(1+j)^{n-k} > \sum_{k=0}^{n} C_k(1+i)^{n-k} = 0.$$

But by (c),

$$\sum_{k=0}^{n} C_k(1+j)^{n-k} = 0,$$

which is a contradiction. (Why were we allowed to assume that $j > i$?) □

Comments About the IRR Uniqueness Theorem II

- Condition (b) requires that $C_0 > 0$. Condition (c) requires that $C_n = -\sum_{k=0}^{n-1} C_k(1+i)^{n-k} = -(1+i)\sum_{k=0}^{n-1} C_k(1+i)^{n-1-k}$, which is negative by conditions (a) and (b). Thus, $C_n < 0$. However, the remaining $C_1, \ldots, C_{n-1}$ have no sign restrictions, other than satisfying (b).
- This theorem guarantees that a typical savings account has a unique IRR, in the following sense. If the savings account is opened with a positive balance $C_0 > 0$, and despite withdrawals and deposits, the interest rate, i, is unchanged and the account is never overdrawn, so $\sum_{k=0}^{p} C_k(1+i)^{p-k} > 0$ for $0 \le p \le n-1$, then there is an IRR, namely i, and it is unique. (Here the quantity C_n is the value of the account, and we are viewing everything through the eyes of the savings institution.)
- This theorem is also valid if the inequality in condition (b) is replaced with $\sum_{k=0}^{p} C_k(1+i)^{p-k} < 0$. (See Problem 2.44 on p. 43.)

2.4 The Rule of 72

The RULE OF 72 is a rule of thumb sometimes used by investors. It states: to calculate the time it takes to double an investment, divide 72 by the annual interest rate expressed as a percentage. For example, if the interest rate is 14%, then, according to the Rule of 72, it takes $72/14 = 5.14$ years to double an investment.

The justification for this rule is based on the following. If interest is compounded continuously at a nominal rate of $i^{(\infty)}$, then the future value at time n is

$$P(n) = P_0 e^{i^{(\infty)} n}.$$

The question we want answered is, at what time $n + T$ is $P(n + T) = 2P(n)$?

The condition $P(n + T) = 2P(n)$ requires that

$$P_0 e^{i^{(\infty)}(n+T)} = 2 P_0 e^{i^{(\infty)} n},$$

or

$$e^{i^{(\infty)} T} = 2,$$

so that

$$T = \frac{\ln 2}{i^{(\infty)}} = \frac{0.693}{i^{(\infty)}}.$$

Thus, if $i^{(\infty)}$ is expressed as a percentage, then the time T needed to double our money is approximately $69.3/i^{(\infty)}$. (Notice that this does not depend on the initial investment nor does it depend on when we start.) Thus, the Rule of 72 should really be called the Rule of 69.3, and it applies only if interest is **compounded continuously**.

However, investments are frequently compound annually, not continuously. So what rule of thumb applies in this case? The answer is, there is no simple rule of thumb like "dividing a particular number by the interest rate". In this case

$$P_n = P_0 (1 + i)^n,$$

so we want to find the N for which

$$P_N = 2P_0,$$

or

$$(1 + i)^N = 2.$$

Solving for N gives

$$N = \frac{\ln 2}{\ln (1 + i)}. \tag{2.8}$$

So the rule of thumb is, divide 0.693 by the natural logarithm of $(1 + i)$. Not exactly a handy rule of thumb!

2.5 Problems

Walking

2.1. Use a spreadsheet to confirm the entries in Table 2.3 on p. 16.

2.2. Tom Kendrick invests $1,000 at a nominal rate of $i^{(1)}$, leaving the interest at the end of each year to compound. At the end of the fourth year he earned $300 in total interest. Determine $i^{(1)}$. Compare your answer to the one you found for Problem 1.1 on p. 10. Which one do you expect to be higher? Explain.

2.3. We invest $1,000 in an account earning 6% per year for 3 years. What is the net present value of our investment if the nominal interest rate is 5%?

2.4. At what nominal interest rate, compounded daily, will money double in 7 years? What is the IRR for this investment?

2.5. How long will it take for an investment of $1,000 to increase to $1,500 at a nominal interest rate of 7% compounded semi-annually? What is the IRR for this investment?

2.6. Tom lends a friend $1,000 on the condition that in 10 years, the friend repays $4,000. What is the IRR for this investment?

2.7. Wendy decides to save for her retirement starting on her 25th birthday. She puts $1,000 a year in an investment that earns 10% a year compounded annually. She does this for 20 years (she is then 45, and has invested $20,000) and then stops adding more money. She then leaves the money invested at 10% annually until she is 65, when she retires. Tom, Wendy's twin brother, does not save for his retirement until his 45th birthday, and then he starts investing a fixed amount each year at 10% per annum for 20 years (at which time both Tom and Wendy are 65). How much does Tom have to invest per year to have the same amount of money as Wendy when she retires?

2.8. Tom buys a stock for $50 and a year later it is worth $100, so the return on Tom's investment for that year is 100%. A year later the stock is worth $50, so the return on Tom's investment for the second year is -50%. Tom claims that the average return on his investment per year over the two year period is $(100 + (-50))/2 = 25\%$. Comment on this claim.

2.9. What is the IRR that corresponds to a simple interest investment rate of 20% over 5 years? Over 4 years? Over 10 years?

2.10 Hugh's wife, Helen Kendrick, buys shares in Cisco for $68. Two months later they are worth $104. What is Helen's IRR?

2.11. Hugh estimates that he needs $1,000,000 when he retires in 15 years. How much must he have in his current retirement account, which earns $8\% a year compounded annually, to reach his goal assuming that he adds no more to his current account?

2.12. An initial amount of $10,000 is invested for 2 years at successive annual interest rates of 10% and 9% compounded annually. Do you think the future value of this investment is different from the future value of $10,000 invested for 2 years at successive annual interest rates of 9% and 10% compounded annually? Justify your answer.

2.13. If the EFF of an investment is 8%, then what is the nominal interest rate, if interest is compounded monthly?

2.14. The National Association of Investors Corporation (NAIC) encourages its members to use a special form (the Stock Selection Guide) to analyze stocks. Part of the form requires the user to convert the projected annual rate of return on an investment over five years from simple to compound interest. The following table is used for this purpose—for example, according to this table five years of simple interest at an annual rate of 10.00% is equivalent to 5 years of compound interest at an annual rate of 8.45%. Confirm that this table is accurate. Is the compound interest row a concave up or a concave down function of the simple interest row?

Simple (%)	10.00	15.00	20.00	25.00	30.00	35.00	40.00
Compound (%)	8.45	11.84	14.87	17.61	20.11	22.42	24.57

2.15. Some banks used to make the promise: "Deposit by the 10th of the month and earn interest from the 1st. Withdraw your money at any time." Two banks, A and B, both compound interest daily at the nominal rate $i^{(360)}$, while only Bank B makes this promise. You deposit an amount in Bank A from the 1st to the 10th of the month, then transfer all your money from Bank A to Bank B on the 10th, and finally transfer all your money from Bank B back to Bank A at the end of the month. You do this every month for a year. Assuming that each month has 30 days, what is the EFF? What is the IRR?

2.16. Use (2.8) on p. 36 to construct a table with the following headings,

Annual Interest Rate	Number of Years	Rule Of

for annual interest rates running from 1% to 30% in increments of 1%. The "Number of Years" column is the number of years it takes for the investment to double at the corresponding annual interest rate. The "Rule Of" column is calculated by multiplying the interest rate (as a percentage) by the number of years, as was the case in the Rule of 72. For what interest rates is there a Rule of 72?

2.17. Use (2.8) on p. 36 to construct a table with the following headings,

Number of Years	Annual Interest Rate	Rule Of

for numbers of years running from 1 to 15 in increments of 1. The "Annual Interest Rate" column is the interest rate (as a percentage) required for the investment to double in the number of years. The "Rule Of" column is calculated by multiplying the interest rate (as a percentage) by the number of years, as was the case in the Rule of 72. For what years is there a Rule of 72?

2.18. Construct a time diagram that represents the cash flows

Years	0	1	2
Cash Flow	$1,000	−$500	−$600

What is the IRR? Give an explanation of these cash flows in terms of everyday experiences, starting with "If I borrow ...."

2.19. Assuming that $P > 0$, construct a time diagram that represents the cash flows

Years	0	1	2	3	4	5
Cash Flow	−P	$0	$0	$1,000	$2,000	$3,000

Find P assuming a 10% nominal interest compounded annually. Give an explanation of these cash flows in terms of everyday experiences, starting with "To be able to withdraw ...."

2.20. Rewrite (2.6) on p. 30 in the form

$$(1+i)^{13} - 16(1+i) + 15 = 0.$$

(a) Find three solutions of this equation by graphing the function $f(x) = x^{13} - 16x + 15$.
(b) Find three solutions of this equation using the bisection method.
(c) Find three solutions of this equation using Newton's Method.
(d) Show that one solution of this equation is $i = 0$. Explain why $i = 0$ is not a solution of (2.5) on p. 29.

2.21. What are the interest rates compounded (a) monthly, (b) semi-annually, and (c) annually that yield the same return as an investment earning 6% interest compounded continuously?

2.22. Consider four different loans for $50,000:

(a) A 10-year loan at 7.5% a year.
(b) A 10-year loan at 10% a year.
(c) A 15-year loan at 7.5% a year.
(d) A 15-year loan at 10% a year.

The monthly payments on the four loans are $593.51, $463.51, $660.76, and $537.71, although not necessarily in that order. The monthly payment for the 10-year loan at 7.5% a year is $593.51. Without using a calculator, match the monthly payments with the loans.

2.23. Find the IRR of a three-year investment of $10,000 that returns the following amounts at the end of each of the three years. What do you conclude?

(a) $400, $500, $10,600.
(b) $500, $600, $10,400.
(c) $600, $400, $10,500.
(d) $600, $500, $10,400.

2.24. Hugh Kendrick is considering three different investments:

(a) One paying 7% compounded quarterly.
(b) One paying 7.1% compounded annually.
(c) One paying 6.9% compounded continuously.

Which has the highest IRR?

2.25. Tom Kendrick is considering two investments with cash flows

Years	0	1	2	3
Cash Flow	−$13,000	$5,000	$6,000	$7,000

What is the IRR? Give an explanation of these cash flows in terms of everyday experiences, starting with "If I borrow ...."

Running

2.26. Show that $P_0(1 + x)^n$ is an increasing, concave up function of x. (See p. 16.)

2.27. Show that $P_0(1 + i)^x$ is an increasing, concave up function of x. (See p. 16.)

2.28. Show, by induction, that for $n > 1$ and $i > 0$,

$$(1 + 2i)^n - 1 > 2\left((1 + i)^n - 1\right).$$

Use this to show that if i, the interest rate per period, is doubled, then the price appreciation from compounding is more than doubled if $n > 1$. What happens if $n = 1$? (See p. 17.)

2.29. Show that for $n > 0$ and $i > 0$,

$$(1+i)^{2n} - 1 > 2\left((1+i)^n - 1\right),$$

by rewriting the inequality as a quadratic in $(1+i)^n$. Use this to show that if n, the number of periods, is doubled, then the principal appreciation from compounding is more than doubled. (See p. 17.)

2.30. We used Fig. 2.2 on p. 17 to suggest that doubling the interest rate more than doubles the principal appreciation, and that doubling the number of periods more than doubles the principal appreciation. Explain how those suggestions could also be made from Fig. 2.1 on p. 17.

2.31. Show that

$$\lim_{m \to \infty} \left(1 + \frac{x}{m}\right)^m = e^x.$$

2.32. Look at (2.2) on p. 19. Based on the fact that $mN = Nm$, do you think that $P_{mN} = P_{Nm}$? If so, prove it. If not, provide a counter-example.

2.33. Show that the EFF of a nominal rate $i^{(m)}$ compounded m times a year is always greater than $i^{(m)}$ for $m > 1$.

2.34. Use induction to sum the geometric series $\sum_{k=1}^{n} x^{k-1}$, that is, prove that

$$\sum_{k=1}^{n} x^{k-1} = \frac{x^n - 1}{x - 1},$$

where $x \neq 1$, for $n = 1, 2, \ldots$ by induction. (See p. 30.)

2.35. Sum the geometric series $\sum_{k=1}^{n} x^{k-1}$, where $x \neq 1$, using the following idea. Let $S_n = \sum_{k=1}^{n} x^{k-1}$. Show that $xS_n - S_n = x^n - 1$. Solve this for S_n. (See p. 30.)

2.36. The geometric mean of the non-negative numbers $a_1, a_2, \ldots, a_n$ is

$$(a_1 a_2 \cdots a_n)^{1/n}.$$

Prove that the IRR of n successive annual investments with rates $i_1, i_2, \ldots, i_n$ compounded $m_1, m_2, \ldots, m_n$ times a year, respectively, is one less than the geometric mean of

$$\left(1 + \frac{i_1}{m_1}\right)^{m_1}, \left(1 + \frac{i_2}{m_2}\right)^{m_2}, \ldots, \left(1 + \frac{i_1}{m_n}\right)^{m_n}.$$

2.37. Show that the IRR that corresponds to a nominal rate $i^{(m)}$ compounded m times a year does not depend on the number of years it is invested.

2.38. Show that the IRR that corresponds to a simple interest investment rate of i depends on the number of years it is invested.

2.39. Show that for a sequence of cash flows over one year, the IRR is the same as the EFF.

2.40. Show that the future value at the end of the first year when invested at a simple annual interest rate of i is the same as the future value at the end of the first year when invested at a compound annual interest rate of i. Show that the future value at the end of the nth year $(n \geq 2)$ when invested at a simple annual interest rate of i is always less than the future value at the end of the nth year when invested at a compound annual interest rate of i.

2.41. An initial amount P_0 is compounded annually for n years at successive annual interest rates of $i_1, i_2, \cdots, i_n$. Show that[11]

$$P_n = P_0 \prod_{j=1}^{n} (1 + i_j).$$

What has this result to do with Problem 2.12 on p. 38?

2.42. The purpose of this problem is to show that the function $(1 + i/t)^t$ is an increasing function of t if $i > 0$.

(a) Consider the function

$$g(x) = \frac{x}{x+1} - \ln(1 + x)$$

for $x > 0$. Show that $g'(x) < 0$. Explain why this leads to $g(0) > g(x)$, that is, if $x > 0$, then

$$\frac{x}{x+1} - \ln(1 + x) < 0.$$

(b) Now consider the function

$$f(x) = (1 + x)^{1/x}$$

for $x > 0$. By considering $f'(x)$ and the inequality from part (a), show that $f(x)$ is a decreasing function of x.

(c) By putting $x = i/t$ in part (b), show that $(1 + i/t)^t$ is an increasing function of t. (Where did you need $i > 0$?)

(d) Explain how part (c) generalizes Theorem 2.2 on p. 20.

---

[11] The product symbol, $\prod_{j=1}^{n} a_j$, is defined by

$$\prod_{j=1}^{n} a_j = a_1 a_2 \cdots a_n.$$

2.43. Is it possible for the cash flows

Time	0	1	2
Cash Flow	C_0	C_1	C_2

to have no IRR if the sequence C_0, C_1, C_2 has exactly one sign change?

2.44. Prove that the IRR Uniqueness Theorem II on p. 34 is also true if the inequality in condition (b) is replaced with $\sum_{k=0}^{p} C_k (1+i)^{p-k} < 0$.

2.45. Give an example of constants $C_0, C_1, \ldots, C_n$ such that there is a unique i that satisfies condition (c) of the IRR Uniqueness Theorem II on p. 34, but not conditions (a) and (b).

2.46. If the rate of change of $P(t)$ with respect to t is proportional to the initial amount $P_0 = P(0)$, that is, if $dP/dt = kP_0$ where k is a positive constant, then show that P is growing at a simple interest rate of k.

2.47. If the rate of change of $P(t)$ with respect to t is proportional to the amount $P(t)$ present at time t, that is, if $dP/dt = kP$, where k is a positive constant, then show that P is growing at an interest rate of k, compounded continuously.

2.48. If the rate of change of $P(t)$ with respect to t is proportional to the square of the amount $P(t)$ present at time t, that is, if $dP/dt = kP^2$, where k is a positive constant, then show that the graph of P has a vertical asymptote. What is the practical significance of this asymptote?

2.49. What compound annual interest rate of i gives the Rule of 72 exactly?

Questions for Review

- How do you calculate compound interest?
- What does compounded continuously mean?
- What is discounting?
- What is the difference between the annual effective rate and the internal rate of return?
- What is a time diagram?
- What is the NPV?
- Why are the IRR Uniqueness Theorems important?
- What is the Rule of 72?
- What is a proof by induction?
- What is a recurrence relation?

3

Inflation and Taxes

In this chapter we discuss two topics that usually reduce the annual effective rate (EFF) of an investment—inflation and taxes. The impact of these two items on any profits should not be ignored.

3.1 Inflation

Inflation changes the purchasing power of money and so behaves like interest but in reverse. For example, if inflation in one year is 10%, then $100 worth of goods at the beginning of the year typically costs $110 at the end. Thus, we are only able to purchase $100/110$ of the goods at the end of the year compared to the beginning. Thus, if the annual inflation rate is i_{inf} (expressed as a decimal), then P_0 at the beginning of a year purchases $P_0/(1 + i_{\text{inf}})$ at the end.

Imagine that we keep our money P_0 under the mattress and that the inflation rate, i_{inf}, is constant. After n years, the purchasing power of our money is reduced to

$$P_n = P_0 \left(\frac{1}{1 + i_{\text{inf}}} \right)^n.$$

Another way of saying this is that P_n future dollars are worth P_0 in TODAY'S DOLLARS.

Theorem 3.1. *The Purchasing Power Theorem.*
If the annual inflation rate is i_{inf}, then the purchasing power of P_0 after n years is reduced to

$$P_n = P_0 \left(\frac{1}{1 + i_{\text{inf}}} \right)^n. \tag{3.1}$$

Comments About the Purchasing Power Theorem

- Table 3.1 shows the impact of inflation on the buying power of $1,000 over time, in five-year intervals, for various annual inflation rates.

Table 3.1. Impact of Inflation on Buying Power of $1,000

Inflation Rate	Years					
	5	10	15	20	25	30
3%	862.61	744.09	641.86	553.68	477.61	411.99
4%	821.93	675.56	555.26	456.39	375.12	308.32
5%	783.53	613.91	481.02	376.89	295.30	231.38
6%	747.26	558.39	417.27	311.80	233.00	174.11
7%	712.99	508.35	362.45	258.42	184.25	131.37
8%	680.58	463.19	315.24	214.55	146.02	99.38
9%	649.93	422.41	274.54	178.43	115.97	75.37

- Figure 3.1 shows the purchasing power of $1,000 as a function of the inflation rate for 10 years (the upper curve) and 20 years (the lower curve).

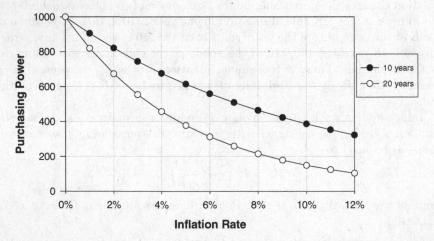

Fig. 3.1. Purchasing power of $1,000 as a function of inflation rate for 10 and 20 years

- Figure 3.2 shows the purchasing power of $1,000 as a function of years for 3% (the upper curve) and 6% (the lower curve).
- Notice, from Figure 3.2, that at a modest 3% inflation rate our purchasing power decreases by 50% in about 23 years. Some parents start saving for

their children's college education shortly after the child's birth. If inflation rates are between 3% and 5% during that time, then they need to save more than twice -as much as the current estimate since increases in the cost of education typically outpace inflation.

- We notice that P_n is directly proportional to P_0 but is a nonlinear function of $i_{\inf}$ and n. Figures 3.1 and 3.2 suggest that the graphs of P_n as a function of i, and P_n as a function of n are both decreasing and concave up.[1] Problems 3.9 and 3.10 on p. 52 ask you to prove these conjectures.

- If $i_{\inf} > 0$, then $\{P_n\}$ is a decreasing sequence for which $\lim_{n \to \infty} P_n = 0$. Thus, during times of inflation, our money buys less and less as time goes by.

- If $i_{\inf} < 0$—this is usually called DEFLATION—then $\{P_n\}$ is an increasing sequence for which $\lim_{n \to \infty} P_n = \infty$. Thus, during times of deflation, our money buys more and more as time goes by.

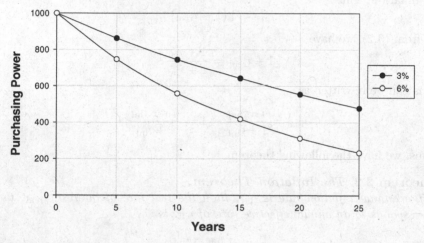

Fig. 3.2. Purchasing power of $1,000 as a function of years for 3% and 6% inflation rates

Example 3.1. Today Amanda Kendrick pays $7 for a movie ticket. In 20 years, how much does she pay in today's dollars if inflation runs at 5%?

Solution. We want to find P_0 in today's dollars if $P_{20} = 7$. From $P_n = P_0(1 + i_{\inf})^{-n}$, we have $P_0 = 7(1 + 0.05)^{20} = 18.57$. Thus, we estimate that a movie ticket costs $18.57 in 20 years due to inflation. $\triangle$

---

[1] Concave up functions are special cases of convex functions, and are widely used in financial circles. (See Problem 3.13 on p. 52.)

Now imagine that we keep our money P_0 in an interest bearing account at an annual effective rate of i_{eff}. If the inflation rate i_{inf} is a constant, then after n years, the purchasing power of our money is

$$P_n = P_0(1 + i_{\text{eff}})^n \left(\frac{1}{1 + i_{\text{inf}}}\right)^n = P_0 \left(\frac{1 + i_{\text{eff}}}{1 + i_{\text{inf}}}\right)^n. \tag{3.2}$$

- If $i_{\text{inf}} > i_{\text{eff}}$, then $\{P_n\}$ is a decreasing sequence for which $\lim_{n \to \infty} P_n = 0$. Thus, as time goes by, our money buys less and less.
- If $i_{\text{inf}} = i_{\text{eff}}$, then $\{P_n\}$ is a constant sequence. As time goes by, our money buys the same amount even though we are earning interest.
- If $i_{\text{inf}} < i_{\text{eff}}$, then $\{P_n\}$ is an increasing sequence for which $\lim_{n \to \infty} P_n = \infty$. Thus, as time goes by, our money buys more and more.

We introduce the REAL RATE OF INTEREST, i_{real}, which is the annual rate at which P_0 grows to P_n when interest is compounded annually at i_{eff} adjusted for inflation. Thus,

$$P_n = P_0(1 + i_{\text{real}})^n.$$

So from (3.2), we have

$$1 + i_{\text{real}} = \frac{1 + i_{\text{eff}}}{1 + i_{\text{inf}}}.$$

This can be rewritten as

$$i_{\text{real}} = \frac{1 + i_{\text{eff}}}{1 + i_{\text{inf}}} - 1 = \frac{i_{\text{eff}} - i_{\text{inf}}}{1 + i_{\text{inf}}}.$$

Thus, we have the following theorem.

Theorem 3.2. *The Inflation Theorem.*
If the annual inflation rate is i_{inf}, then the real rate of interest, i_{real}, that corresponds to an annual effective rate of i_{eff}, is

$$i_{\text{real}} = \frac{i_{\text{eff}} - i_{\text{inf}}}{1 + i_{\text{inf}}}. \tag{3.3}$$

Solving (3.3) for i_{eff} gives

$$i_{\text{eff}} = i_{\text{inf}} + i_{\text{real}} + i_{\text{inf}} i_{\text{real}}.$$

3.2 Consumer Price Index (CPI)

The Consumer Price Index (CPI) is a measure of the average price over time of a fixed market basket of goods and services bought by consumers for day-to-day living. The CPI for the United States is the broadest, most comprehensive index and is often quoted as the source for the "rate of inflation".[2]

---

[2] Some people feel that the CPI is an upper limit for the rate of inflation.

Table 3.2 shows the average annual CPI, and how it has changed from 1913 to 2005, with the CPI of 100 occurring in mid-1983.[3]

Table 3.2. Average Annual CPI

Year	Index	%	Year	Index	%	Year	Index	%
1913	9.9		1944	17.6	1.7	1975	53.8	9.1
1914	10.0	1.0	1945	18.0	2.3	1976	56.9	5.8
1915	10.1	1.0	1946	19.5	8.3	1977	60.6	6.5
1916	10.9	7.9	1947	22.3	14.4	1978	65.2	7.6
1917	12.8	17.4	1948	24.1	8.1	1979	72.6	11.3
1918	15.1	18.0	1949	23.8	−1.2	1980	82.4	13.5
1919	17.3	14.6	1950	24.1	1.3	1981	90.9	10.3
1920	20.0	15.6	1951	26.0	7.9	1982	96.5	6.2
1921	17.9	−10.5	1952	26.5	1.9	1983	99.6	3.2
1922	16.8	−6.1	1953	26.7	0.8	1984	103.9	4.3
1923	17.1	1.8	1954	·26.9	0.7	1985	107.6	3.6
1924	17.1	0.0	1955	26.8	−0.4	1986	109.6	1.9
1925	17.5	2.3	1956	27.2	1.5	1987	113.6	3.6
1926	17.7	1.1	1957	28.1	3.3	1988	118.3	4.1
1927	17.4	−1.7	1958	28.9	2.8	1989	124.0	4.8
1928	17.1	−1.7	1959	29.1	0.7	1990	130.7	5.4
1929	17.1	0.0	1960	29.6	1.7	1991	136.2	4.2
1930	16.7	−2.3	1961	29.9	1.0	1992	140.3	3.0
1931	15.2	−9.0	1962	30.2	1.0	1993	144.5	3.0
1932	13.7	−9.9	1963	30.6	1.3	1994	148.2	2.6
1933	13.0	−5.1	1964	31.0	1.3	1995	152.4	2.8
1934	13.4	3.1	1965	31.5	1.6	1996	156.9	3.0
1935	13.7	2.2	1966	32.4	2.9	1997	160.5	2.3
1936	13.9	1.5	1967	33.4	3.1	1998	163.0	1.6
1937	14.4	3.6	1968	34.8	4.2	1999	166.6	2.2
1938	14.1	−2.1	1969	36.7	5.5	2000	172.2	3.4
1939	13.9	−1.4	1970	38.8	5.7	2001	177.1	2.8
1940	14.0	0.7	1971	40.5	4.4	2002	179.9	1.6
1941	14.7	5.0	1972	41.8	3.2	2003	184.0	2.3
1942	16.3	10.9	1973	44.4	6.2	2004	188.9	2.7
1943	17.3	6.1	1974	49.3	11.0	2005	195.3	3.4

The % column represents the percentage change in the index from one year to the next. For example, the percentage change of 3.4 in 2005 is calculated from

$$\frac{\text{Index in 2005} - \text{Index in 2004}}{\text{Index in 2004}} = \frac{195.3 - 188.9}{188.9} = 0.0339,$$

which is 3.4% to one decimal place.

---

[3] ftp://ftp.bls.gov/pub/special.requests/CPI/cpiai.txt, accessed March 17, 2006.

From the table, we see that in 1970 the index was 38.8, and by 2005 it had grown to 195.3. This suggests that goods that cost \$38.80 in 1970 cost \$195.30 in 2005. Thus, after 35 years an amount P_0 has its purchasing power reduced to $P_{35} = (38.8/195.3)P_0$, and so the inflation rate, i_{inf}, for those 35 years must satisfy

$$\frac{38.8}{195.3}P_0 = P_0 \left(\frac{1}{1 + i_{\text{inf}}}\right)^{35},$$

that is,

$$i_{\text{inf}} = \left(\frac{195.3}{38.8}\right)^{1/35} - 1 = 0.047.$$

So inflation has been about 5% per year over those 35 years.

3.3 Personal Taxes

If we keep our money P_0 in an interest bearing account at an annual effective rate of i_{eff}, then after one year we have $P_0(1 + i_{\text{eff}})$. Thus, our pre-tax profit that year is $i_{\text{eff}}P_0$. This profit incurs taxes, and if t is our tax rate (expressed as a decimal), then we pay $i_{\text{eff}}P_0 t$ in taxes.[4] Thus, the actual after-tax amount we have available to re-invest the second year is

$$P_1 = P_0(1 + i_{\text{eff}}) - i_{\text{eff}}P_0 t = P_0(1 + i_{\text{eff}} - i_{\text{eff}}t).$$

One year later this has grown to $P_1(1 + i_{\text{eff}})$, pre-tax, and so after taxes we have

$$P_2 = P_1(1 + i_{\text{eff}}) - i_{\text{eff}}P_1 t = P_1(1 + i_{\text{eff}} - i_{\text{eff}}t).$$

Continuing in this way, we find that

$$P_{k+1} = P_k(1 + i_{\text{eff}}) - i_{\text{eff}}P_k t = P_k(1 + i_{\text{eff}} - i_{\text{eff}}t).$$

This is the same recurrence relation that we found in the proof of Theorem 2.1 on p. 15, which has solution

$$P_n = P_0(1 + i_{\text{eff}} - i_{\text{eff}}t)^n.$$

Thus, we have the following theorem.

---

[4] In the United States, as of July 2006, the tax rates on earned income depend on marital status and taxable income. The rates are: 15%, 25%, 28%, and 35%. The tax rates on unearned income (such as investments) are different from those on earned income (such as salary).

Theorem 3.3. *The Tax Theorem.*
If the annual effective rate is i_{eff} and the annual tax rate, t, is constant, then the after-tax future value of P_0 after n years is

$$P_n = P_0(1 + i_{eff} - i_{eff}t)^n. \tag{3.4}$$

We now introduce the AFTER-TAX RATE OF INTEREST, i_{tax}, which is the after-tax annual rate at which P_0 grows to P_n when compounded annually at that rate. Thus,

$$P_n = P_0(1 + i_{tax})^n,$$

and so from (3.4), we have

$$1 + i_{tax} = 1 + i_{eff} - i_{eff}t.$$

This can be rewritten as

$$i_{tax} = i_{eff} - i_{eff}t = i_{eff}(1 - t). \tag{3.5}$$

3.4 Problems

Walking

3.1. Confirm that the entries in Table 3.1 on p. 46 are correct.

3.2. In what years did the United States experience deflation?

3.3. What is the rate of inflation for the period 1980–2005? The period 1990–2005? The period 2000–2005?

3.4. In 1974, Helen Kendrick received 10% interest compounded annually on her savings account. Should she be pleased? What was the real rate of interest?

3.5. On average, the cost of a college textbook is about $75. How much does it cost 25 years from now (in today's dollars)—when the children of current freshmen are themselves freshmen—if inflation runs at 5%?

3.6. In January 1970, Hugh Kendrick bought a new Toyota Corolla for $2,000. In January 2000, he bought one for $14,000. What annual rate of inflation does this correspond to?

3.7. Some investments are tax-exempt. The following table shows that, for a person in the 15% taxable income bracket, an annual effective rate of 2.35% on a taxable investment is equivalent to an annual effective rate of 2% on a tax-exempt investment. Complete this table.

Tax Exempt Return	Tax Rate			
	15%	25%	28%	35%
2%	2.35%	2.67%	2.78%	3.08%
3%				
4%				
5%				
6%				
7%				
8%				
9%				
10%				

Running

3.8. A rule of thumb used to estimate the real interest rate is $i_{real} \approx i_{eff} - i_{inf}$. Show that this estimate is always too high during times of inflation and too low during times of deflation.

3.9. Show that

$$f(x) = P_0 \left(\frac{1}{1 + i_{inf}} \right)^x$$

is a decreasing, concave up function of x for $i_{inf} > 0$. (See p. 47.)

3.10. Show that

$$g(x) = P_0 \left(\frac{1}{1 + x} \right)^n$$

is a decreasing, concave up function of x. (See p. 47.)

3.11. Some people think, incorrectly, that the inflation equation is not (3.2) on p. 48, but is given by $P_0(1 - i_{inf})^n$. Show that these people predict a value that is always lower than the correct one when $-1 < i_{inf} < 1$.

3.12. If P_0 is placed in an interest bearing account at an annual effective rate of i_{eff}, if the annual inflation rate is i_{inf}, and if the annual tax rate is t, then what is the after-tax after-inflation rate of interest?

3.13. A function $f(x)$ on an interval I is said to be convex on I if for every $p \in (0,1)$ and every $x, y \in I$, the function $f(x)$ satisfies[5]

$$f(px + (1-p)y) \leq pf(x) + (1-p)f(y).$$

(a) Show that the function $f(x) = x$ is convex.
(b) Show that the function $f(x) = |x|$ is convex.
(c) Show that the function $f(x) = x^2$ is convex.

---

[5] The notation $a \in I$, means that a is an element or member of I. If I is not specified it is assumed to be $(-\infty, \infty)$.

3.14. Give an example of a function that is not convex.

3.15. Consider the function

$$f(x) = \begin{cases} x & \text{if } x \le 1, \\ 2 - x & \text{if } x > 1. \end{cases}$$

Show that $f(x)$ is convex on $-\infty < x \le 1$ and convex on $1 < x < \infty$ but not convex on $-\infty < x < \infty$.

3.16. If $f(x)$ is convex on an interval I, if $p_i \ge 0$ $(i = 1, 2, \ldots, n)$ where $p_1 + p_2 + \cdots p_n = 1$, and if $x_1, x_2, \cdots, x_n \in I$, then show that

$$f\left(\sum_{i=1}^{n} p_i x_i\right) \le \sum_{i=1}^{n} p_i f(x_i).$$

[Hint: Use induction.]

3.17. Suppose that $f(x)$ is defined on the interval I. If $x_1 < x_2 < x_3$ are in I and $f(x_1) < f(x_2)$ and $f(x_3) < f(x_2)$, then show that $f(x)$ is not convex on I.

Questions for Review

- What is inflation?
- What is deflation?
- What is meant by the expression "the purchasing power of money"? How is it calculated?
- What is the real rate of interest?
- What is the Consumer Price Index?
- What is the after-tax rate of interest?

Annuities

An ANNUITY is a constant amount of money paid at regular intervals, called periods. When the payments are made at the end of the period, the annuity is called an ORDINARY ANNUITY. When payments are made at the beginning of the period, the annuity is called an ANNUITY DUE.

Annuities occur in a variety of different settings. For example, someone who is saving for retirement by investing a constant amount of money at the end of every month in an account that pays a fixed interest rate is creating an ordinary annuity. Someone who wins the lottery and who has selected the annuity payout option is paid a constant amount of money immediately and every year thereafter for a fixed number of years—an annuity due.

It is important that annuities be thoroughly understood—they are fundamental to the remaining chapters.

4.1 An Ordinary Annuity

We invest $1,000 at the end of each year at 10% annual interest for 5 years. At the end of the first year we earn no interest because we have just deposited $1,000. At the end of the second year we earn 10% of $1,000, namely, $100, and we deposit $1,000, totaling $2,100. At the end of the third year we earn 10% of $2,100, namely, $210, and we deposit $1,000, totaling $3,310. We continue doing this for 5 years. Table 4.1 shows the details. (Check the calculations in this table using a calculator or a spreadsheet program, and fill in the missing entries.)

We now derive the general formula for this process. We let

P be the AMOUNT invested at the end of every period,
n be the TOTAL NUMBER OF PERIODS,
P_n be the FUTURE VALUE of the annuity at the end of the nth period,
m be the NUMBER OF PERIODS PER YEAR,
$i^{(m)}$ be the NOMINAL RATE (annual interest rate), expressed as a decimal,
i be the INTEREST RATE PER INTEREST PERIOD.

Table 4.1. Ordinary Annuity

Year	Year's Beginning Principal	Interest	Year's End Investment	Amount
1	$0.00	$0.00	$1,000.00	$1,000.00
2	$1,000.00	$100.00	$1,000.00	$2,100.00
3	$2,100.00	$210.00	$1,000.00	$3,310.00
4				
5	$4,641.00	$464.10	$1,000.00	$6,105.10

The interest rate per period is $i = i^{(m)}/m$.

We want to find a formula for the future value P_n, and we do this by looking at $n = 1$, $n = 2$, and so on, hoping to see a pattern. Using this notation, we rewrite Table 4.1 symbolically in spreadsheet format as Table 4.2, which is explained as follows.

Table 4.2. Ordinary Annuity—Spreadsheet Format

Period	Period's Beginning Principal	Interest	Period's End Investment	Amount
1	0	0	P	$0 + 0 + P = P_1$
2	P_1	iP_1	P	$P_1 + iP_1 + P = P_2$
3	P_2	iP_2	P	$P_2 + iP_2 + P = P_3$
4	P_3	iP_3	P	$P_3 + iP_3 + P = P_4$
5	P_4	iP_4	P	$P_4 + iP_4 + P = P_5$

At the end of period 1 we have made 1 payment, and our future value is

$$P_1 = P.$$

At the end of period 2 we have made 2 payments, and our future value is

$$P_2 = P_1(1 + i) + P = P(1 + i) + P.$$

At the end of period 3 we have made 3 payments, and our future value is

$$P_3 = P_2(1 + i) + P = P(1 + i)^2 + P(1 + i) + P.$$

At the end of period 4 we have made 4 payments, and our future value is

$$P_4 = P_3(1 + i) + P = P(1 + i)^3 + P(1 + i)^2 + P(1 + i) + P.$$

This suggests that at the end of period n we have made n payments, and our future value is

$$P_n = P(1 + i)^{n-1} + P(1 + i)^{n-2} + \cdots + P(1 + i) + P.$$

Using the geometric series

$$1 + x + \cdots + x^{n-1} = \frac{x^n - 1}{x - 1}, \ x \neq 1,$$

we can rewrite this conjecture in closed form[1]

$$P_n = P\left((1+i)^{n-1} + (1+i)^{n-2} + \cdots + (1+i) + 1\right) = P\frac{(1+i)^n - 1}{i}.$$

This suggests the following theorem.

Theorem 4.1. *The Future Value of an Ordinary Annuity Theorem.* *If we invest P at the end of every period for n periods (where there are m periods per year) at a nominal rate of $i^{(m)}$ (expressed as a decimal), then P_n, the future value of the annuity after n periods, is*

$$P_n = P\sum_{k=1}^{n} (1+i)^{k-1} = P\frac{(1+i)^n - 1}{i}, \tag{4.1}$$

where $i = i^{(m)}/m$.

Proof. We can prove this conjecture in at least two different ways: either using mathematical induction or using recurrence relations.

We first prove it using mathematical induction. We already know that (4.1) is true for $n = 1$. We assume that it is true for $n = k$, that is,

$$P_k = P\frac{(1+i)^k - 1}{i}.$$

and we must show that it is true for $n = k + 1$. Now,

$$P_{k+1} = P_k(1+i) + P,$$

so

$$P_{k+1} = P\frac{(1+i)^k - 1}{i}(1+i) + P = P\frac{(1+i)^{k+1} - 1}{i},$$

which shows that (4.1) is true for $n = k + 1$. This concludes the proof by mathematical induction.

We now prove it using recurrence relations. We know that $P_1 = P$ and that for $k > 0$,

$$P_{k+1} = P_k(1+i) + P,$$

so if we multiply this by $1/(1+i)^{k+1}$, then we can write it as

---

[1] Loosely speaking, an expression is in "closed form" if it does not contain terms such as "$\cdots$".

$$P_{k+1}\frac{1}{(1+i)^{k+1}} = P_k\frac{1}{(1+i)^k} + P\frac{1}{(1+i)^{k+1}}.$$

Summing this from $k = 1$ to $k = n - 1$ (where $n > 1$) gives

$$\sum_{k=1}^{n-1} P_{k+1}\frac{1}{(1+i)^{k+1}} = \sum_{k=1}^{n-1} P_k\frac{1}{(1+i)^k} + P\sum_{k=1}^{n-1} \frac{1}{(1+i)^{k+1}},$$

or by canceling the common terms on both sides of this equation,

$$P_n\frac{1}{(1+i)^n} = P\frac{1}{1+i} + P\sum_{k=1}^{n-1} \frac{1}{(1+i)^{k+1}},$$

that is,

$$P_n = P\sum_{k=0}^{n-1} \frac{(1+i)^n}{(1+i)^{k+1}}.$$

However,

$$\sum_{k=0}^{n-1} \frac{(1+i)^n}{(1+i)^{k+1}} = 1 + (1+i) + \cdots + (1+i)^{n-1},$$

which is a geometric series that can be written in the form

$$\sum_{k=0}^{n-1} \frac{(1+i)^n}{(1+i)^{k+1}} = \frac{(1+i)^n - 1}{i}.$$

Thus,

$$P_n = P\frac{(1+i)^n - 1}{i},$$

which is (4.1). This concludes the proof using recurrence relations.   □

Comments About the Future Value of an Ordinary Annuity Theorem

- The general situation is represented by Fig. 4.1.

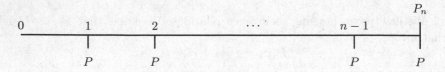

Fig. 4.1. Time diagram of an ordinary annuity

- Notice that (4.1), namely,

$$P_n = P(1+i)^{n-1} + P(1+i)^{n-2} + \cdots + P(1+i) + P,$$

  is consistent with Fig. 4.1 because the future value of the first payment P (after n periods) is $P(1+i)^{n-1}$, the future value of the second payment P (after n periods) is $P(1+i)^{n-2}$, and so on.
- Table 4.3 shows the future value of an ordinary annuity of \$1 for different annually-compounded interest rates and different numbers of years.

Table 4.3. Future Value of an Ordinary Annual Annuity of \$1

Interest Rate	Years					
	5	10	15	20	25	30
3%	5.309	11.464	18.599	26.870	36.459	47.575
4%	5.416	12.006	20.024	29.778	41.646	56.085
5%	5.526	12.578	21.579	33.066	47.727	66.439
6%	5.637	13.181	23.276	36.786	54.865	79.058
7%	5.751	13.816	25.129	40.995	63.249	94.461
8%	5.867	14.487	27.152	45.762	73.106	113.283
9%	5.985	15.193	29.361	51.160	84.701	136.308
10%	6.105	15.937	31.772	57.275	98.347	164.494
11%	6.228	16.722	34.405	64.203	114.413	199.021
12%	6.353	17.549	37.280	72.052	133.334	241.333
13%	6.480	18.420	40.417	80.947	155.620	293.199
14%	6.610	19.337	43.842	91.025	181.871	356.787
15%	6.742	20.304	47.580	102.444	212.793	434.745

- We notice that P_n is a function of the three variables P, n, and i. It is directly proportional to P—which is why we selected $P = 1$ in Table 4.3—but nonlinear in i and n.
- The dependence of P_n on i is seen in Fig. 4.2, which shows the future value of an ordinary annuity of \$1 as a function of i for 10 years (the lower curve) and 20 years (the upper curve) with annual compounding. Notice that both curves appear to be increasing and concave up. In Problem 4.18 on p. 73 you are asked to prove this.
- The dependence of P_n on n is seen in Fig. 4.3, which shows the future value of an ordinary annuity of \$1 as a function of n for 5% interest (the lower curve) and 10% interest (the upper curve) compounded annually. Notice that both curves appear to be increasing and concave up. In Problem 4.19 on p. 73 you are asked to prove this.

Example 4.1. Helen and Hugh Kendrick decide to give Amanda, their 20-year old daughter, an extended birthday gift. They deposit \$2,000 in her name on

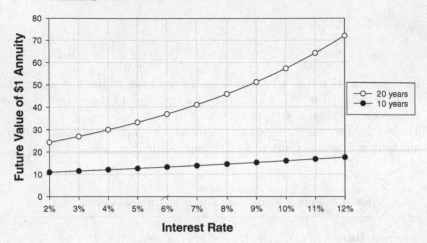

Fig. 4.2. Future Value of $1 annuity for 10 and 20 years with annual compounding

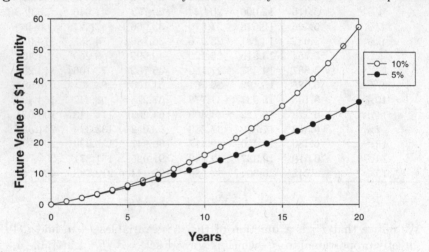

Fig. 4.3. Future Value of $1 annuity at 5% and 10% interest compounded annually

her next five birthdays at a nominal rate of 8% compounded annually, and then they stop the annual deposits. Amanda leaves the accumulated amount invested at 8%. How much does Amanda have when she turns 65?

Solution. In this case Helen and Hugh make five deposits, so $n = 5$, and at the end of the fifth year Amanda has

$$P_5 = 2000 \frac{(1 + 0.08)^5 - 1}{0.08} = \$11{,}733.20.$$

This is now invested for 40 years at 8%, giving

$$11733.20 \, (1 + 0.08)^{40} = \$254{,}898.16. \quad \triangle$$

Example 4.2. If, under the circumstances of Example 4.1, a rival institution offered the Kendricks a 10% interest rate (instead of 8%), then how much does Amanda have when she turns 65?

Solution. At the end of the fifth year she has

$$P_5 = 2000\frac{(1+0.10)^5 - 1}{0.10} = \$12,210.20.$$

This is now invested for 40 years at 10%, giving

$$12210.20\,(1+0.10)^{40} = \$552,624.56.$$

$\triangle$

Notice that a relatively small change in interest rates, in this case from 8% to 10%, can make a huge difference in the value of the investment after a number of years, in this case a difference of about $300,000 on an initial investment of $10,000.

Example 4.3. Tom Kendrick is repaying a student loan with a nominal interest of 5% at $100 a month. He misses three successive payments but makes the fourth. He then pays an additional lump sum[2] to make up for the three missed payments. How much is the lump sum? How much extra interest does he pay as a result of his late payments?

Solution. Tom's scheduled cash flows are represented by Fig. 4.4, where P_4 is the total value of the four payments.

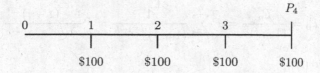

Fig. 4.4. Time diagram of Tom's scheduled cash flows

We recognize this as an ordinary annuity with $P = \$100$, $n = 4$, and $i = 0.05/12$, so

$$P_4 = 100\frac{(1+\frac{0.05}{12})^4 - 1}{\frac{0.05}{12}} = \$402.51.$$

Thus, the lump sum is $P_4 - 100 = \$302.51$. The $302.51 - \$300.00 = \$2.51$ is the extra interest for the late payments. An alternative way to think of this problem is represented by Fig. 4.5, where L is the lump sum owed due to the missed payments.

---

[2] In practice, there might be an additional charge for missing the payments.

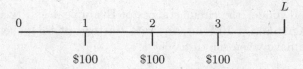

Fig. 4.5. Time diagram of Tom's debt

The future value of the first missed payment is $100\left(1+0.05/12\right)^3$, of the second missed payment is $100\left(1+0.05/12\right)^2$, and of the final missed payment is $100\left(1+0.05/12\right)$. Thus,

$$L = 100\left(1+\frac{0.05}{12}\right)^3 + 100\left(1+\frac{0.05}{12}\right)^2 + 100\left(1+\frac{0.05}{12}\right) = \$302.51.$$

$\triangle$

The present value of P_n is P_0, where $P_n = P_0\left(1+i\right)^n$, so the present value of an ordinary annuity is

$$P_0 = P_n(1+i)^{-n} = P\frac{(1+i)^n - 1}{i}(1+i)^{-n} = P\frac{1-(1+i)^{-n}}{i}.$$

We thus have the following theorem.

Theorem 4.2. *The Present Value of an Ordinary Annuity Theorem. If we invest P at the end of every period for n periods (where there are m periods per year) at a nominal rate of $i^{(m)}$ (expressed as a decimal), then P_0, the present value of the annuity, is*

$$P_0 = \frac{P_n}{(1+i)^n} = P\sum_{k=1}^{n}\frac{1}{(1+i)^{n+1-k}} = P\frac{1-(1+i)^{-n}}{i}, \tag{4.2}$$

where $i = i^{(m)}/m$.

Comments About the Present Value of an Ordinary Annuity Theorem

- Notice that (4.2), namely,

$$P_0 = P\sum_{k=1}^{n}\frac{1}{(1+i)^{n+1-k}} = \frac{P}{1+i} + \frac{P}{(1+i)^2} + \cdots + \frac{P}{(1+i)^n},$$

is represented by Fig. 4.6.

This is interpreted as receiving an initial loan of P_0 and repaying P over n payment periods—called amortizing a loan. Given any positive P, i, and n (where n is an integer), we use (4.2) to compute P_0. In other words, we can answer questions like, "If I can afford to pay P a month for n months,

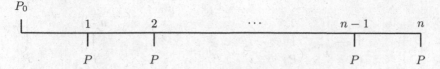

Fig. 4.6. Time diagram of a simple loan

then how much can I borrow at a monthly interest rate of i?" However, in general and for a very subtle reason, we cannot answer the question, "If I want to borrow P_0 at a monthly interest rate of i and can afford to pay P a month, then how long does it take to repay the loan?" To answer this question, we have to solve (4.2), namely,

$$P_0 = P\frac{1 - (1 + i)^{-n}}{i},$$

for n, obtaining

$$n = \frac{\ln\left(\frac{P}{P - iP_0}\right)}{\ln(1 + i)}.$$

However, in general the left-hand side is an integer, whereas the right-hand side is not. We return to this type of problem in Chap. 6.

Example 4.4. The Kendricks are thinking of buying a new house by amortizing a loan. They can afford to pay \$700 a month for 30 years. If the current interest rate is 8%, then how much can they borrow?

Solution. From (4.2), with $P = 700$, $m = 12$, $i = 0.08/12$, and $n = 30 \times 12 = 360$, we find that

$$P_0 = 700\frac{1 - \left(1 + \frac{0.08}{12}\right)^{-360}}{\frac{0.08}{12}} = \$95{,}398.45.$$

Thus, they can borrow about \$95,000. $\triangle$

A SINKING FUND is an account, earning a nominal interest rate of $i^{(m)}$, into which a constant amount of money P is deposited regularly m times a year (often monthly) with a view to accumulating a targeted amount F on a specified date n periods in the future. If the deposits are made at the end of each period, including the final period, then the time diagram is Fig. 4.7.

This is just an ordinary annuity, so from (4.1), we have

$$F = P\frac{(1 + i)^n - 1}{i}, \tag{4.3}$$

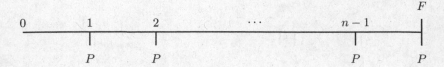

Fig. 4.7. Time diagram of a sinking fund

where $i = i^{(m)}/m$. Usually the question with a sinking fund is, "How much must we deposit every period to have a targeted final figure at a specified time?" In other words, find P in terms of F, n, and i. Thus, from (4.3),

$$P = F\frac{i}{(1+i)^n - 1}. \tag{4.4}$$

Example 4.5. Hugh Kendrick expects to replace his car in 5 years. He estimates that he needs \$20,000. How much per month must he put into a sinking fund at a nominal rate of 6% compounded monthly to replace his car?

Solution. Here $F = 20000$, $n = 5 \times 12 = 60$, and $i = 0.06/12 = 0.005$, so

$$P = 20000\frac{0.005}{(1.005)^{60} - 1} = 286.656,$$

namely, \$286.66 a month. $\triangle$

We can do two back-of-the-envelope, order-of-magnitude calculations to see whether \$286.66 a month is in the right ballpark.

- First, intuition suggests that the total deposited is less than the final amount because most deposits earn interest. The total deposited is \$286.66 × 60 = \$17,199.60, which is less than \$20,000. This intuitive calculation suggests that for $n > 1$,

$$nP < F.$$

But is our intuition correct?
For $nP < F$ to be true we need, from (4.3),

$$n < \frac{(1+i)^n - 1}{i},$$

or

$$(1+i)^n - 1 - ni > 0.$$

We prove this by induction on $n > 1$. With $n = 2$, the left-hand side of this inequality becomes $(1+i)^2 - 1 - 2i = i^2 > 0$, so it is true for $n = 2$. We assume that it is true for $n = k$, that is,

$$(1+i)^k - 1 - ki > 0,$$

and consider the left-hand side with $n = k + 1$, namely, $(1 + i)^{k+1} - 1 - (k + 1)i$. Now,

$$\begin{aligned}
(1 + i)^{k+1} - 1 - (k + 1)i &= (1 + i)^k(1 + i) - 1 - (k + 1)i \\
&> (1 + ki)(1 + i) - 1 - (k + 1)i \\
&= ki^2,
\end{aligned}$$

which is positive. Thus, our intuition is correct.

- Second, intuition suggests that the total deposited is greater than the present value of the final amount because each new deposit only earns interest from the time it is made to the time of the final deposit, whereas the present value of the final amount is computed assuming that the investment earns interest immediately. The present value of \$20,000 is $\$20,000(1 + 0.005)^{-60} = \$14,827.44$, which is less than $\$286.66 \times 60 = \$17,199.60$. This intuitive calculation suggests that for $n > 0$,

$$F(1 + i)^{-n} < nP.$$

But is our intuition correct? For this to be true we need, from (4.3),

$$\frac{(1 + i)^n - 1}{i}(1 + i)^{-n} < n,$$

or

$$(1 + i)^n(1 - ni) < 1.$$

This is clearly true if $1 - ni \leq 0$, so we prove this by induction on $n > 0$ assuming that $1 - ni > 0$, that is, $n < 1/i$.

With $n = 1$, the left-hand side of this inequality becomes $(1 + i)(1 - i) = 1 - i^2 < 1$, so it is true for $n = 1$. We assume that it is true for $n = k$, that is,

$$(1 + i)^k(1 - ki) < 1,$$

and consider the left-hand side with $n = k + 1$, namely, $(1 + i)^{k+1}(1 - (k + 1)i)$.

Now,

$$\begin{aligned}
(1 + i)^{k+1}(1 - (k + 1)i) &= (1 + i)^k(1 + i)(1 - ki - i) \\
&< \tfrac{1}{1-ki}(1 + i)(1 - ki - i) \\
&= 1 + i - \tfrac{i(1+i)}{1-ki} \\
&= 1 - \tfrac{i^2(1+k)}{1-ki} \\
&< 1.
\end{aligned}$$

Thus, our intuition is correct again. (Where did we use the fact that $n < 1/i$?)

Thus, we can always check the numerical value of P using the following result.

Theorem 4.3. *For a sinking fund, the periodic payment P given by (4.4),*

$$P = F\frac{i}{(1+i)^n - 1},$$

must satisfy the inequality

$$\frac{F}{n}(1+i)^{-n} < P < \frac{F}{n},$$

for $n > 1$.

Example 4.6. Jana Carmel, a coworker of Hugh Kendrick, is 35 years old and wants to have \$1 million when she retires at age 65. If she earns 10% interest compounded annually on her savings, then how much must she save per year assuming no inflation? Assuming inflation at 5% per year?

Solution. Here $F = 10^6$, $n = 65 - 35 = 30$, and $i = 0.1/1 = 0.1$, so we know that $10^6/30 \times (1.1)^{-30} < P < 10^6/30$, that is, \$1,910.29 $< P <$ \$33,333.33. In fact,

$$P = 10^6 \frac{0.1}{(1.1)^{30} - 1} = \$6,079.25.$$

Thus, Jana must save \$6,079.25 a year for 30 years to become a before-inflation millionaire.

If inflation runs at 5% per year, then by the Purchasing Power Theorem on p. 45, the purchasing power of 10^6 dollars 30 years from now is $F = 10^6 (1 + 0.05)^{30}$, so

$$P = 10^6 (1 + 0.05)^{30} \frac{0.1}{(1.1)^{30} - 1} = \$26,274.16.$$

Thus, Jana should save \$26,274.16 a year for 30 years to become an after-inflation millionaire. The moral of the story is: inflation matters. $\triangle$

4.2 An Annuity Due

Now let us assume that we invest \$1,000 at the **beginning** of each year at 10% annual interest for 5 years. At the end of the first year we earn 10% of \$1,000, namely, \$100, totaling \$1,100. At the beginning of the second year we deposit an additional \$1,000, totaling \$2,100, so at end of the second year we earn 10% of \$2,100, namely, \$210, totaling \$2,310. We continue doing this for 5 years. Table 4.4 shows the details. (Check the calculations in this table using a calculator or a spreadsheet program, and fill in the missing entries.)

Table 4.4. Annuity Due

Year	Year's Beginning		Year's End	
	Investment	Principal	Interest	Amount
1	$1,000.00	$1,000.00	$100.00	$1,100.00
2	$1,000.00	$2,100.00	$210.00	$2,310.00
3	$1,000.00	$3,310.00	$331.00	$3,641.00
4				
5	$1,000.00	$6,105.10	$610.51	$6,715.61

We now derive the general formula for this process. We let

P be the AMOUNT invested at the beginning of every period,
n be the TOTAL NUMBER OF PERIODS,
P_n be the FUTURE VALUE of the annuity at the end of the nth period,
m be the NUMBER OF PERIODS PER YEAR,
$i^{(m)}$ be the NOMINAL RATE (annual interest rate), expressed as a decimal,
i be the INTEREST RATE PER INTEREST PERIOD.

The interest rate per period is $i = i^{(m)}/m$.
The general situation is represented by Fig. 4.8.

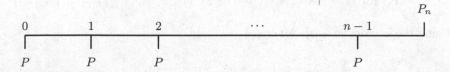

Fig. 4.8. Time diagram of an annuity due

We want to find a formula for the future value P_n, and we do this by looking
at $n = 1$, $n = 2$, and so on, hoping to see a pattern. Using this notation, we
rewrite Table 4.4 symbolically in spreadsheet format as Table 4.5, which is
explained as follows.

Table 4.5. Annuities Due—Spreadsheet Format

Period	Period's Beginning		Period's End	
	Investment	Principal	Interest	Amount
1	P	P	iP	$P + iP = P_1$
2	P	$P + P_1$	$i(P + P_1)$	$P + P_1 + i(P + P_1) = P_2$
3	P	$P + P_2$	$i(P + P_2)$	$P + P_2 + i(P + P_2) = P_3$
4	P	$P + P_3$	$i(P + P_3)$	$P + P_3 + i(P + P_3) = P_4$
5	P	$P + P_4$	$i(P + P_4)$	$P + P_4 + i(P + P_4) = P_5$

If we make only 1 payment, then at the end of period 1 our future value is

$$P_1 = P(1+i).$$

If we make 2 payments, then at the end of period 2 our future value is

$$P_2 = (P + P_1)(1+i) = P(1+i) + P(1+i)^2.$$

If we make 3 payments, then at the end of period 3 our future value is

$$P_3 = (P + P_2)(1+i) = P(1+i) + P(1+i)^2 + P(1+i)^3.$$

If we make 4 payments, then at the end of period 4 our future value is

$$P_4 = (P + P_3)(1+i) = P(1+i) + P(1+i)^2 + P(1+i)^3 + P(1+i)^4.$$

This suggests that if we make n payments, then at the end of period n our future value is

$$P_n = P(1+i) + P(1+i)^2 + \cdots + P(1+i)^{n-1} + P(1+i)^n,$$

which we write in the form

$$P_n = P\left(1 + (1+i) + (1+i)^2 + \cdots + (1+i)^{n-2} + (1+i)^{n-1}\right)(1+i).$$

Using the geometric series for $x \neq 1$,

$$1 + x + \cdots + x^{n-1} = \frac{x^n - 1}{x - 1},$$

we rewrite this conjecture in closed form

$$P_n = P\frac{(1+i)^n - 1}{i}(1+i).$$

Thus, we have the following theorem.

Theorem 4.4. *The Future Value of an Annuity Due Theorem.*
If we invest P at the beginning of every period for n periods (where there are m periods per year) at a nominal rate of $i^{(m)}$ (expressed as a decimal), then P_n, the future value of the annuity after n periods, is

$$P_n = P\frac{(1+i)^n - 1}{i}(1+i), \tag{4.5}$$

where $i = i^{(m)}/m$.

The proof of this theorem is developed in Problem 4.21 on p. 73.

Comments About the Future Value of an Annuity Due Theorem

- The future value of an annuity due is always greater than the future value of an ordinary annuity with the same P, n, m, and i because it is $(1 + i)$ times the future value of an ordinary annuity.
- The future value of an annuity due is always increasing and concave up as a function of i because it is $(1 + i)$ times the future value of an increasing and concave up ordinary annuity. (See Problem 4.23 on p. 73.)

Example 4.7. Repeat Example 4.1 on p. 59 for the case when all payments are made one year earlier.

Solution. At the end of the fifth year Amanda has

$$P_5 = 2000 \frac{(1 + 0.08)^5 - 1}{0.08} (1 + 0.08) = \$12,671.86.$$

This is now invested for 40 years at 8%, giving

$$12671.86 \, (1 + 0.08)^{40} = \$275,290.09,$$

a difference of over \$20,000 from the result of Example 4.1. △

Example 4.8. Repeat Example 4.2 on p. 61 for the case when all payments are made one year earlier.

Solution. At the end of the fifth year Amanda has

$$P_5 = 2000 \frac{(1 + 0.10)^5 - 1}{0.10} (1 + 0.10) = \$13,431.22.$$

This is now invested for 40 years at 10%, giving

$$13431.22 \, (1 + 0.10)^{40} = \$607,887.02,$$

a difference of over \$55,000 from the result of Example 4.2. △

The present value of P_n is P_0, where $P_n = P_0(1 + i)^n$. From this we may derive the following theorem.

Theorem 4.5. *The Present Value of an Annuity Due Theorem.*
If we invest P at the beginning of every period for n periods (where there are m periods per year) at a nominal rate of $i^{(m)}$ (expressed as a decimal), then P_0, the present value of the annuity, is

$$P_0 = \frac{P_n}{(1 + i)^n} = P \frac{1 - (1 + i)^{-n}}{i} (1 + i), \tag{4.6}$$

where $i = i^{(m)}/m$.

In Problem 4.22 on p. 73 you are asked to prove this theorem.

Example 4.9. Jana Carmel wins the Arizona Lottery. She receives $207,850 immediately and at the end of each of the following 19 years for a total of 20 payments.

(a) The State of Arizona buys an annuity to cover this. If the nominal interest rate is 7.06%, then how much does the state pay for this annuity?
(b) If Jana pays 28% in taxes, then what is her after-taxes yearly payout?
(c) If Jana pays 28% in taxes and inflation averages 5% a year, then what is her final payment worth, after taxes, in today's dollars?

Solution.

(a) Because Jana received the first payment immediately, this is an annuity due where $P = 207850$, $m = 1$, $i^{(1)} = 0.0706$, and $n = 20$, so the present value (which is what the state pays) is

$$P_0 = 207850 \frac{1 - (1 + 0.0706)^{-20}}{0.0706} (1 + 0.0706) = \$2,346,471.02.$$

(b) After taxes, $207,850 becomes $207850 \times (1 - 0.28) = \$149,652$.
(c) After taxes and after 19 years inflation (remember, her final payment is made at the end of the 19th year) this is worth $149652 (1/1.05)^{19} = \$59,222.38$ in today's dollars, which is less than one third of the original yearly payout.

$\triangle$

4.3 Perpetuities

In case you think that your intuition is infallible—so it is really not worth proving anything—consider the following situation.

Example 4.10. Which would you prefer to receive: $10,000 at the end of every year forever or $200,000 now if both are invested at 5% compounded annually?

Solution. The way to compare these two choices is to calculate the present value, P_0, of the first choice and compare that to $200,000. The present value is

$$P_0 = 10000(1 + 0.05)^{-1} + 10000(1 + 0.05)^{-2} + \cdots + 10000(1 + 0.05)^{-n} + \cdots.$$

This can be written as

$$P_0 = 10000(1 + 0.05)^{-1} \left(1 + (1 + 0.05)^{-1} + \cdots + (1 + 0.05)^{-n+1} + \cdots\right),$$

or

$$P_0 = 10000(1+0.05)^{-1} \sum_{n=0}^{\infty} \left(\frac{1}{1+0.05}\right)^n.$$

The series is the infinite geometric series, $\sum_{n=0}^{\infty} x^n$, which converges to $1/(1-x)$ if $|x| < 1$. In our case $x = 1/1.05$, so

$$P_0 = 10000(1.05)^{-1} \frac{1}{1 - \frac{1}{1.05}} = \frac{10000}{0.05} = \$200,000.$$

Thus, you should have no preference between the two propositions. To some this is counter-intuitive because they think that \$10,000 a year forever is the better choice. It isn't. In fact, if instead of receiving \$10,000 a year forever, you only receive \$10,000 a year for a finite number of years, no matter how many, the \$200,000 now is the better deal. $\triangle$

An annuity that goes on forever is called a PERPETUITY. The present value, P_0, of a perpetuity of payments in the amount of P per payment period at interest i per payment period, is given by

$$P_0 = P(1+i)^{-1} + P(1+i)^{-2} + \cdots + P(1+i)^{-n} + \cdots,$$

that is,

$$P_0 = P(1+i)^{-1} \sum_{n=0}^{\infty} \left(\frac{1}{1+i}\right)^n = P(1+i)^{-1} \frac{1}{1 - \frac{1}{1+i}} = \frac{P}{i}.$$

We thus have the following result.

Theorem 4.6. *The present value, P_0, of a perpetuity of payments in the amount of P per payment period at interest i per payment period, is*

$$P_0 = \frac{P}{i}.$$

4.4 Problems

Walking

4.1. Use a spreadsheet program to confirm the entries in Table 4.3 on p. 59.

4.2. Use Table 4.2 on p. 56 as a template to construct a spreadsheet that solves Examples 4.1 on p. 59 and 4.2 on p. 61.

4.3. The following is a quotation from The Arizona Daily Star (February 27, 2000): "If a person saved \$50 a month for 20 years and earned 8% a year over the entire period, that person would end up with \$29,451.02." Is this statement true?

4.4. The Department of Mathematics at the University of Arizona wants to set up a fund to assist graduate students with travel expenses to conferences. The department estimates that it needs to withdraw $1,000 from the fund at the end of each year, forever. If the fund earns 7% interest per annum, then how much money needs to be placed in the fund at the beginning?

4.5. Repeat Problem 4.4 taking into account inflation at 3% a year.

4.6. The cost of a four-year college education at a public university is expected to be $7,000 a year in 18 years. How much money should be invested now at 7% so that the balance of the account after 18 years covers the cost of a college education?

4.7. A person deposits $50 at the end of every month for the next 10 years in an account earning 9% compounded monthly. What is the balance of the account at the end of the 10 years?

4.8. Repeat Problem 4.7 under the assumption that the deposits are made at the beginning of each month rather than at the end of each month.

4.9. A college education is expected to cost $20,000 per year in 18 years. How much money should be deposited at the end of each year for 18 years in an account earning 10% compounded monthly so that the balance of the account after 18 years covers the cost of a college education?

4.10. Repeat Problem 4.9 under the assumption that the deposits are made at the beginning of each year rather than at the end of each year.

4.11. In five years, a person plans to buy a car that is currently valued at $16,000. The expected inflation rate is 3% per year. How much money should that person invest at the end of each month for 60 months in an account earning 9% compounded monthly so that the balance of the account at the end of the five years covers the cost of the car?

4.12. The cost of a college education is currently $15,000 per year. The expected annual inflation rate over the next five years is 3%. How much money should be deposited at the end of each year for five years in an account earning 10% compounded quarterly so that the balance of the account after five years covers the cost of a college education?

4.13. A person pays $100,000 for an investment that promises to pay $15,000 per year at the end of each year for 10 years. What is the rate of return on the investment?

4.14. Repeat Problem 4.13 under the assumption that the payments are made at the beginning of each year rather than at the end of each year.

4.15. A person pays $50,000 for an investment that promises to pay a fixed amount of money at the end of each year for 15 years. If the rate of return is 8%, then what is the amount of the annual payments?

4.16. Repeat Problem 4.15 under the assumption that the payments are made at the beginning of each year rather than at the end of each year.

4.17. An investment promises to pay $50 per year in perpetuity. If the investor's tax rate is 30% and if the investor can deposit the payments into an account earning 10% compounded monthly, then what is the present value of the perpetuity?

Running

4.18. Show that
$$f(x) = P_0 \frac{(1+x)^n - 1}{x}$$
is an increasing, concave up function of x. What is $\lim_{x \to 0+} f(x)$? (See p. 59.)

4.19. Show that
$$g(x) = P_0 \frac{(1+i)^x - 1}{i}$$
is an increasing, concave up function of x. What is $\lim_{i \to 0+} g(x)$? (See p. 59.)

4.20. Prove Theorem 4.3 on p. 66 by induction.

4.21. Prove (4.5) on p. 68.

4.22. Prove (4.6) on p. 69.

4.23. Show that for $x > 0$, if $f(x)$ is positive, increasing, and concave up, then so is $g(x) = (1+x)f(x)$. (See p. 69.)

4.24. It is claimed that an annuity due can be thought of as one initial payment and $n - 1$ ordinary annuity payments. Is this true?

Questions for Review

- What is an annuity?
- What is the difference between an ordinary annuity and an annuity due?
- How do you calculate the future value of an ordinary annuity? An annuity due?
- How do you calculate the present value of an ordinary annuity? An annuity due?
- What is a sinking fund?
- What is a perpetuity?

5

Loans and Risks

Many students obtain STUDENT LOANS. These loans may be issued by individuals, businesses (as part of an employee's benefit program), or by the government. A common type of student loan is a Stafford loan, which is guaranteed by the federal government. Typical conditions for a Stafford loan are that the repayment period of the loan is at most 10 years, the minimum monthly payment is $50, and there is no prepayment penalty (this last condition is important because many real estate loans, for example, charge a very high penalty for paying off the loan early). These loans are currently at a fixed rate, and there is a late fee for late payments. An important feature of these loans is that the student does not begin to repay the loan until 6 months after completing the academic program or leaving school.

HOME LOANS are available through banks, savings and loans, and mortgage brokers. Home loans can be at a fixed or adjustable rate.

- Fixed rate loans are usually given for terms of 15 or 30 years. Usually, the interest rate for a 30-year loan is higher than that for a 15-year loan. In this case the borrower pays an added premium (a higher interest rate) for the privilege of paying off the loan over a longer period of time.

- An adjustable rate loan offers a fixed rate for an initial period of time (typically 1 to 5 years). At the end of this initial period the rate may be adjusted, usually in relation to some government index, such as the Treasury bill rate. Generally, the initial rate for an adjustable rate loan is lower than the comparable rate for a fixed rate loan. The borrower is given this consideration to compensate for the uncertainty of future rates. At the end of the initial period, periodic reviews of the loan are made, and the rate is adjusted according to the index being used. Usually there is a "cap"—a maximum amount that the rate may be increased at any time.

Home loans often carry a prepayment penalty. If homeowners wish to pay off their loans early in order to refinance at a lower rate, then a prepayment penalty may make this a bad decision. A prepayment penalty may be as much as six months interest on 80% of the remaining balance.

ʙᴜsɪɴᴇss ʟᴏᴀɴs are available from many banks, individuals, and local, state, or federal government agencies. In many cases the seller of a business may "carry back" a loan to the buyer. In these cases there may be great flexibility in the structure of the loan. For example, a common practice is to defer payment of the principal and interest until the new owner has been in business for a fixed period of time. This is an example of a concession by the owner. In other cases there may be a "balloon" payment. In these cases the buyer makes periodic payments until a fixed date, at which time the remainder of the principal is due. This is an example of a concession by the buyer.

Borrowers often have many choices when deciding how to finance purchases. The choices vary with respect to the term of the loan, the interest rate charged for the loan, whether or not payments include principal, and whether or not the loan is secured. For example, business loans tend to have higher rates than home loans, real estate loans for non-owner-occupied properties generally have higher interest rates than for owner-occupied properties, and fixed rate loans generally have higher initial interest rates than adjustable rate loans. The differences in choices are associated with differences in the associated risks. In this and other financial contexts, the term risk is synonymous with uncertainty.

One source of risk is interest rate risk. The more quickly a loan is repaid, the lower the impact of unexpected increases in interest rates on the lender. Therefore, short-term loans and loans that include principal repayments are less risky than long-term loans and loans that do not include principal repayments.

Another source of risk is default risk. Default risk is the risk that a borrower fails to make interest or principal payments when promised. Again, the more quickly a loan is repaid, the less risky the loan for the lender; if a borrower defaults on a loan, then the amount of the outstanding principal is lower for short-term loans that include principal repayments. Also, secured loans are less risky than unsecured loans. If a borrower fails to repay a loan that is secured by collateral, then the lender can take possession of the collateral. However, this is not the case with unsecured loans. These differences in risk result in differences in the rates charged for the various loans; higher rates are charged for riskier loans.

However, even loans with the same interest rate don't necessarily have the same total interest. We now look at four different ways of repaying a loan, all based on the fact that we borrow the same amount at the same interest rate.

Loan 1: Consider an annuity in which we pay $1,000 at the end of each year for 5 years. Current interest rates are 8%. We know from (4.2) on p. 62 that we can borrow

$$P_0 = P\frac{1 - (1 + i)^{-n}}{i} = 1000\frac{1 - (1 + 0.08)^{-5}}{0.08} = \$3{,}992.71.$$

Table 5.1 shows the details of this loan repayment.

Table 5.1. Amortization

Year	Year's Start Remaining Principal	Interest Owed	Payment	Year's End Interest Paid	Principal Paid	Remaining Principal
1	$3,992.71	$319.42	$1,000.00	$319.42	$680.58	$3,312.13
2	$3,312.13	$264.97	$1,000.00	$264.97	$735.03	$2,577.10
3	$2,577.10	$206.17	$1,000.00	$206.17	$793.83	$1,783.27
4	$1,783.27	$142.66	$1,000.00	$142.66	$857.34	$925.93
5	$925.93	$74.07	$1,000.00	$74.07	$925.93	$0.00

The total interest paid is $1,007.29. This is typical of an **amortization**, which was discussed in Section 4.1 and is discussed fully in Chap. 6. An amortized loan is one for which constant principal and interest payments are made at regular intervals until the loan is repaid.

The IRR, i_A, for these transactions satisfies

$$3992.71 - \frac{1000}{1 + i_A} - \frac{1000}{(1 + i_A)^2} - \frac{1000}{(1 + i_A)^3} - \frac{1000}{(1 + i_A)^4} - \frac{1000}{(1 + i_A)^5} = 0,$$

or

$$3992.71x^5 - 1000x^4 - 1000x^3 - 1000x^2 - 1000x - 1000 = 0.$$

where $x = 1 + i_A$. This polynomial in x has only one change of sign and so has at most one positive solution, namely, $x = 1 + i_A = 1.08$, so $i_A = 0.08$.

The remaining three loans are also for $3,992.71 at 8%, but the method of repayment differs in each case.

Loan 2: Another institution offers to lend us the same $3,992.71 at 8% over five years, but we are only required to pay the interest at the end of each year and the original principal at the end of the fifth year. Table 5.2 shows the details of the loan repayment.

Table 5.2. Bond

Year	Year's Start Remaining Principal	Interest Owed	Payment	Year's End Interest Paid	Principal Paid	Remaining Principal
1	$3,992.71	$319.42	$319.42	$319.42	$0.00	$3,992.71
2	$3,992.71	$319.42	$319.42	$319.42	$0.00	$3,992.71
3	$3,992.71	$319.42	$319.42	$319.42	$0.00	$3,992.71
4	$3,992.71	$319.42	$319.42	$319.42	$0.00	$3,992.71
5	$3,992.71	$319.42	$4,312.13	$319.42	$3,992.71	$0.00

The total interest paid is \$1,597.08. This is typical of **bonds**, which are discussed in Chap. 8. A bond is a loan that an investor makes to a governmental agency or corporation for which the borrower pays the lender a fixed amount at regular intervals until the last payment, at which time the principal is repaid.

The IRR, i_B, for these transactions satisfies

$$3992.71 - \frac{319.42}{1+i_B} - \frac{319.42}{(1+i_B)^2} - \frac{319.42}{(1+i_B)^3} - \frac{319.42}{(1+i_B)^4} - \frac{4312.13}{(1+i_B)^5} = 0.$$

This polynomial in $1+i_B$ has only one change of sign and so has at most one positive solution, namely, $1+i_B = 1.08$, so $i_B = 0.08$.

Loan 3: Another institution offers to lend us the same \$3,992.71 at 8% over five years, but we are only required to pay the principal and interest at the end of the fifth year. Table 5.3 shows the details of this loan repayment.

Table 5.3. Zero Coupon Bond

Year	Year's Start Remaining Principal	Interest Owed	Payment	Year's End Interest Paid	Principal Paid	Remaining Principal
1	\$3,992.71	\$319.42	\$0.00	\$0.00	\$0.00	\$4,312.13
2	\$4,312.13	\$344.97	\$0.00	\$0.00	\$0.00	\$4,657.10
3	\$4,657.10	\$372.57	\$0.00	\$0.00	\$0.00	\$5,029.67
4	\$5,029.67	\$402.37	\$0.00	\$0.00	\$0.00	\$5,432.04
5	\$5,432.04	\$434.56	\$5,866.60	\$1,873.89	\$3,992.71	\$0.00

Note that for example, the remaining principal at the start of year 3 (\$4,657.10) includes the remaining principal at the start of year 2 (\$4,312.13) plus the interest accrued during that year (\$344.97). The interest for year 3 is calculated from the remaining principal at the start of that year (\$4,657.10). Thus, interest is computed upon interest.

The total interest paid is \$1,873.89. This is typical of **zero coupon bonds**, which are discussed in Chap. 8. A zero coupon bond is a bond for which there are no regular payments. Thus, the only payment made is a lump sum payment at the end of the loan period.

The IRR, i_Z, for these transactions satisfies

$$3992.71 - \frac{5866.60}{(1+i_Z)^5} = 0,$$

or

$$i_Z = \left(\frac{5866.60}{3992.71}\right)^{1/5} - 1 = 0.08.$$

Loan 4: Another institution offers to lend us the same $3,992.71 at 8%, but we are only required to pay $500 a year until the debt is repaid, and no interest is charged the first year. Table 5.4 shows the details of this loan repayment.

Table **5.4.** Credit Card

Year	Year's Start Remaining Principal	Interest Owed	Payment	Year's End Interest Paid	Principal Paid	Remaining Principal
1	$3,992.71	$0.00	$500.00	$0.00	$500.00	$3,492.71
2	$3,492.71	$279.42	$500.00	$279.42	$220.58	$3,272.13
3	$3,272.13	$261.77	$500.00	$261.77	$238.23	$3,033.90
4	$3,033.90	$242.71	$500.00	$242.71	$257.29	$2,776.61
5	$2,776.61	$222.13	$500.00	$222.13	$277.87	$2,498.74
6	$2,498.74	$199.90	$500.00	$199.90	$300.10	$2,198.64
7	$2,198.64	$175.89	$500.00	$175.89	$324.11	$1,874.53
8	$1,874.53	$149.96	$500.00	$149.96	$350.04	$1,524.49
9	$1,524.49	$121.96	$500.00	$121.96	$378.04	$1,146.45
10	$1,146.45	$91.72	$500.00	$91.72	$408.28	$738.16
11	$738.16	$59.05	$500.00	$59.05	$440.95	$297.22
12	$297.22	$23.78	$321.00	$23.78	$297.22	$0.00

The total interest paid is $1828.29, and it takes 12 years to pay off the debt. This is typical of a **credit card**. Credit cards are discussed in Chap. 7. A credit card is issued by a financial institution and regular payments are made on the outstanding balance of the loan. The minimum payment is the greater of a fixed rate times the outstanding balance and a fixed dollar amount.

The IRR, i_C, for these transactions satisfies

$$3992.71 - 500 - \sum_{k=1}^{10} \frac{500}{(1 + i_C)^k} - \frac{321}{(1 + i_C)^{11}} = 0.$$

This has only one change of sign and so has at most one positive solution, namely, $1 + i_C = 1.08$, so $i_C = 0.08$.

5.1 Problems

Walking

5.1. Helen Kendrick wishes to borrow $50,000 when the interest rate is 7.5%. She is only willing to pay the interest at the end of each year and the original principal at the end of the fourth year. Construct the payment schedule for this loan.

5.2. Hugh Kendrick has a 30-year home loan for $100,000 at 7%. After 15 years, he wishes to pay off the loan in order to refinance at a lower rate. The loan has a prepayment penalty of six months interest on 80% of the remaining balance of the loan.

(a) How much is the remaining balance on Hugh's loan?
(b) If Hugh can refinance with a 15-year loan at 6% with no other costs, then should he do it?
(c) If Hugh can refinance the loan over 15 years with no other costs, then what interest rate would make it worthwhile?

5.3. Helen Kendrick wishes to borrow money when the interest rate is 8.5%, and she is willing to pay $5,000 at the end of each year for four years.

(a) How much money can she borrow?
(b) Create a table similar to Table 5.1 on p. 77 for this loan.

5.4. Helen Kendrick wishes to borrow $50,000 when the interest rate is 7.5%. She needs cash and is only willing to pay interest and principal at the end of the fourth year. Construct the payment schedule for this loan.

5.5. Helen Kendrick wishes to borrow $50,000 when the interest rate is 7.5%. She is only willing to pay $3,000 interest for the first three years and then pay off the loan at the end of the fourth year. Construct the payment schedule for this loan.

5.6. Consider two four-year loans—one for $50,000 and one for $100,000—both at 7.5%. Treating these loans as amortization loans, is the monthly payment of the second loan twice the monthly payment of the first loan? Is the amount of interest paid during any given month for the second loan twice the interest paid during the same month for the first loan?

Running

5.7. The following is a partially completed table similar to Table 5.1 on p. 77 for an amortized loan.

Year	Year's Start Remaining Principal	Int. Owed	Payment	Year's End Int. Paid	Princ. Paid	Remaining Principal
1	$10,000.00		$2,820.12			
2						
3						
4						$0.00

(a) What is the interest rate?
(b) Complete the table.

5.8. Under what circumstances is an adjustable rate loan preferable to a fixed rate loan?

5.9. Compare the four different ways of repaying a loan discussed in this chapter. Noting that they all have an IRR of 8%, are there situations where one of these loans may be better than the others?

5.10. Hugh Kendrick wishes to borrow $10,000. He must choose between two fixed rate five-year loans: one at 7% with no prepayment penalty, and one at 6.75% with a 1% prepayment penalty. Under what circumstances is the second loan preferable to the first loan?

5.11. Referring to Problem 5.10, how much is Hugh's final payment if he chooses the loan with the prepayment penalty and pays off the loan immediately after his third payment? Will he have paid more or less in total than he would had he taken the other loan?

Questions for Review

- How does risk impact the interest rate charged to a borrower?
- What are some of the risks associated with loans?

6

Amortization

The largest investment that most people make is buying a house. However, it is unusual to buy a house with cash. Most people borrow the money from a company that issues mortgages. The same process used to repay mortgages is frequently used to repay student loans and car loans. In Chap. 5 we mentioned amortization. In this chapter we study the process in detail.

6.1 Amortization Tables

Initially an amount of money (the initial principal or term amount) is borrowed at a specified annual interest rate, and principal and interest payments are made at constant periodic intervals for a certain time until the debt is repaid—this is called "amortizing the loan". Lending companies have tables—called AMORTIZATION TABLES—in which the various payments are tabulated by interest rate and term. For example, Table 6.1 is part of one such table where the nominal interest rate is 10% and monthly payments are made. Thus, according to this table if we borrow $10,000 for 10 years (at 10%), then our monthly payment is $132.16.

With this information, we complete Table 6.2—called an AMORTIZATION SCHEDULE—which shows the first five months of the loan repayment. At the end of the first month we pay $132.16. Of this, $0.10/12 \times 10000 = \$83.33$ is the interest owed, and $132.16 - \$83.33 = \$48.83$ goes to reducing the principal to $\$10,000.00 - \$48.83 = \$9,951.17$. (Check the calculations in this table using a calculator or a spreadsheet program, and fill in the missing entries.)

Table 6.1. Part of an Amortization Table

Term Amount	10 years	11 years	12 years	13 years	14 years
$10,000	$132.16	$125.20	$119.51	$114.79	$ 110.83
$11,000	$145.37	$137.72	$131.46	$126.27	$121.91

Table 6.2. Amortization Schedule

Month	Monthly Payment	Interest Paid	Principal Repaid	Outstanding Principal
0	$0.00	$0.00	$0.00	$10,000.00
1	$132.16	$83.33	$48.83	$9,951.17
2	$132.16	$82.93	$49.23	$9,901.94
3	$132.16	$82.52	$49.64	$9,852.30
4				
5	$132.16	$81.69	$50.47	$9,751.77

Notice that Table 6.2 depends critically on knowing Table 6.1. But how do we know Table 6.1, and more importantly, how do we know that it is accurate?

In this section we show how these loan payments are computed. We let

P_0 be the INITIAL PRINCIPAL (term amount) that has to be repaid,

P_n be the OUTSTANDING PRINCIPAL (principal remaining to be paid) at the end of the nth payment period,

M be the PERIODIC PAYMENT,

N be the total number of PAYMENT PERIODS,

m be the NUMBER OF PERIODIC PAYMENTS PER YEAR,

$i^{(m)}$ be the ANNUAL INTEREST RATE, expressed as a decimal, and

i be the PERIODIC INTEREST RATE,

so $i = i^{(m)}/m$.

Figure 6.1 shows a time diagram for this.

Fig. 6.1. Time diagram of a mortgage

Using this notation, we rewrite Table 6.2 symbolically in spreadsheet format as Table 6.3, which is explained as follows.

At the end of the first payment period ($n = 1$) we make our periodic payment of M, but the interest owed is iP_0, so the outstanding principal is

$$P_1 = P_0 - M + iP_0 = P_0(1 + i) - M.$$

At the end of the second payment period ($n = 2$) the interest owed is iP_1, so the outstanding principal is

$$P_2 = P_1 - M + iP_1 = P_1(1 + i) - M = (P_0(1 + i) - M)(1 + i) - M,$$

Table 6.3. Amortization Schedule—Spreadsheet Format

Period	Periodic Payment	Interest Paid	Principal Repaid	Outstanding Principal
0				P_0
1	M	iP_0	$M - iP_0$	$P_0 - (M - iP_0) = P_1$
2	M	iP_1	$M - iP_1$	$P_1 - (M - iP_1) = P_2$
3	M	iP_2	$M - iP_2$	$P_2 - (M - iP_2) = P_3$
4	M	iP_3	$M - iP_3$	$P_3 - (M - iP_3) = P_4$
5	M	iP_4	$M - iP_4$	$P_4 - (M - iP_4) = P_5$

which reduces to

$$P_2 = P_0(1+i)^2 - M(1 + (1+i)).$$

At the end of the third payment period ($n = 3$) the interest owed is iP_2, so the outstanding principal is

$$P_3 = P_2 - M + iP_2,$$

which reduces to

$$P_3 = P_0(1+i)^3 - M\left(1 + (1+i) + (1+i)^2\right).$$

This suggests that P_n, the outstanding principal at the end of the nth payment period, is

$$P_n = P_0(1+i)^n - M\left(1 + (1+i) + \cdots + (1+i)^{n-1}\right).$$

We rewrite this conjecture, using the fact that for

$$1 + x + \cdots + x^{n-1} = \frac{x^n - 1}{x - 1} \text{ for } x \neq 1,$$

in the form[1]

$$P_n = P_0(1+i)^n - M\frac{(1+i)^n - 1}{i}, \tag{6.1}$$

or

$$P_n = \left(P_0 - \frac{M}{i}\right)(1+i)^n + \frac{M}{i}.$$

---

[1] Equation (6.1) can be interpreted as the future value of the principal, $P_0(1+i)^n$, less the future value of the annuity payments, $M((1+i)^n-)/i$.

Thus, we have the following theorem.

Theorem 6.1. *The Amortization Theorem.*
If P_0 is the initial principal borrowed at a nominal interest rate of $i^{(m)}$ and if periodic payments of M are made m times a year, then P_n, the outstanding principal after n payment periods, is

$$P_n = \left(P_0 - \frac{M}{i}\right)(1+i)^n + \frac{M}{i}, \tag{6.2}$$

where $i = i^{(m)}/m$.

Proof. We can prove this conjecture either by mathematical induction or by recurrence relations. We prove it using recurrence relations and leave the proof by induction for Problem 6.12 on p. 98.

We know that

$$P_{k+1} = P_k(1+i) - M,$$

so if we multiply this by $1/(1+i)^{k+1}$, then we can write it as

$$P_{k+1}\frac{1}{(1+i)^{k+1}} = P_k\frac{1}{(1+i)^k} - M\frac{1}{(1+i)^{k+1}}.$$

Summing this from $k = 0$ to $k = n - 1$ gives

$$\sum_{k=0}^{n-1} P_{k+1}\frac{1}{(1+i)^{k+1}} = \sum_{k=0}^{n-1} P_k\frac{1}{(1+i)^k} - M\sum_{k=0}^{n-1}\frac{1}{(1+i)^{k+1}},$$

or by canceling the common terms on both sides of this equation,

$$P_n\frac{1}{(1+i)^n} = P_0 - M\sum_{k=0}^{n-1}\frac{1}{(1+i)^{k+1}},$$

that is,

$$P_n = P_0(1+i)^n - M(1+i)^n\sum_{k=0}^{n-1}\frac{1}{(1+i)^{k+1}}.$$

However,

$$(1+i)^n\sum_{k=0}^{n-1}\frac{1}{(1+i)^{k+1}} = 1 + (1+i) + \cdots + (1+i)^{n-1} = \frac{(1+i)^n - 1}{i}.$$

Thus,

$$P_n = P_0(1+i)^n - M\frac{(1+i)^n - 1}{i},$$

which is (6.2). This concludes the proof using recurrence relations. $\square$

Because

$$P_n = \left(P_0 - \frac{M}{i}\right)(1+i)^n + \frac{M}{i}$$

and because $\{(1+i)^n\}$ is an increasing sequence, that is, $(1+i)^n < (1+i)^{n+1}$, we see that $P_n < P_{n+1}$ if $P_0 - M/i > 0$ and that $P_n > P_{n+1}$ if $P_0 - M/i < 0$. We thus have the following intuitively obvious result.

Theorem 6.2. *Let P_n be given by (6.2).*

(a) *If $M > iP_0$, then $P_n > P_{n+1}$, that is, if the periodic payment exceeds the initial periodic interest owed, then the outstanding principal decreases.*

(b) *If $M = iP_0$, then $P_n = P_{n+1}$, that is, if the periodic payment exactly equals the initial periodic interest owed, then the outstanding principal is constant.*

(c) *If $M < iP_0$, then $P_n < P_{n+1}$, that is, if the periodic payment is less than the initial periodic interest owed, then the outstanding principal increases and is never paid off.*

From now on we concentrate on the case $M > iP_0$, which also implies that $M > iP_n$ (why?), and we rewrite (6.2) in the form

$$P_n = \frac{M}{i} - \left(\frac{M}{i} - P_0\right)(1+i)^n.$$

In this form we see that mathematically $\lim_{n\to\infty} P_n = -\infty$. In view of this fact and that $\{P_n\}$ is a decreasing sequence for which $P_0 > 0$, there must be an integer N for which $P_{N-1} > 0$ and $P_N \le 0$. This N represents the number of payment periods required to repay the loan. Thus, N must satisfy

$$\frac{M}{i} - \left(\frac{M}{i} - P_0\right)(1+i)^{N-1} > 0$$

and

$$\frac{M}{i} - \left(\frac{M}{i} - P_0\right)(1+i)^N \le 0,$$

so

$$\left(\frac{M}{i} - P_0\right)(1+i)^{N-1} < \frac{M}{i} \le \left(\frac{M}{i} - P_0\right)(1+i)^N.$$

This can be rewritten in the form

$$\frac{M}{M - iP_0} \le (1+i)^N < \frac{M}{M - iP_0}(1+i), \tag{6.3}$$

so that N must satisfy

$$\frac{\ln\left(\frac{M}{M-iP_0}\right)}{\ln\left(1+i\right)} \leq N < \frac{\ln\left(\left(\frac{M}{M-iP_0}\right)(1+i)\right)}{\ln\left(1+i\right)}.$$

A natural question to ask is whether more than one integer N can satisfy (6.3). Intuitively we expect not, and this is seen by rewriting the last inequality as

$$\frac{\ln\left(\frac{M}{M-iP_0}\right)}{\ln\left(1+i\right)} \leq N < \frac{\ln\left(\frac{M}{M-iP_0}\right)}{\ln\left(1+i\right)} + 1. \tag{6.4}$$

6.2 Periodic Payments

If we finish repaying the loan **exactly** at the end of the Nth payment period, then we have $P_N = 0$ and we can then calculate the periodic payment M. In (6.2) we set $n = N$ and $P_N = 0$ and solve for M, finding

$$M = \frac{iP_0\left(1+i\right)^N}{\left(1+i\right)^N - 1},$$

so the **periodic payments** are

$$M = \frac{iP_0}{1 - \left(1+i\right)^{-N}}.$$

Theorem 6.3. The Periodic Payment Theorem.
If P_0 is the initial principal borrowed at a nominal interest rate of $i^{(m)}$ for N payment periods, then M, the payment that is made m times a year, is

$$M = \frac{iP_0}{1 - \left(1+i\right)^{-N}}, \tag{6.5}$$

where $i = i^{(m)}/m$.

Comments About the Periodic Payment Theorem

- We notice that M is a function of the three variables P_0, i, and N. The quantity M is directly proportional to P_0. Table 6.4 and Fig. 6.2 show M as a function of i. Notice that all the curves appear to be increasing and concave up. In Problem 6.15 on p. 99 you are asked to prove this. Table 6.5 and Fig. 6.3 show M as a function of N. Notice that all the curves appear to be decreasing and concave up. In Problem 6.16 on p. 100 you are asked to prove this.

Table 6.4. Monthly payments as a function of the annual interest rate for 5-, 15-, and 30-year loans in the amount of $10,000

Interest Rate	5 years	15 years	30 years
1%	$170.94	$59.85	$32.16
2%	$175.28	$64.35	$36.96
3%	$179.69	$69.06	$42.16
4%	$184.17	$73.97	$47.74
5%	$188.71	$79.08	$53.68
6%	$193.33	$84.39	$59.96
7%	$198.01	$89.88	$66.53
8%	$202.76	$95.57	$73.38
9%	$207.58	$101.43	$80.46
10%	$212.47	$107.46	$87.76
11%	$217.42	$113.66	$95.23
12%	$222.44	$120.02	$102.86

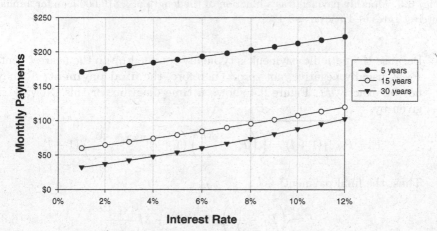

Fig. 6.2. Monthly payments as a function of the annual interest rate for 5-, 15-, and 30-year loans in the amount of $10,000

Table 6.5. Monthly payments as a function of the length of a $10,000 loan for annual interest rates of 4%, 8%, and 12%

Length of Loan	4%	8%	12%
5	$184.17	$202.76	$222.44
10	$101.25	$121.33	$143.47
15	$73.97	$95.57	$120.02
20	$60.60	$83.64	$110.11
25	$52.78	$77.18	$105.32
30	$47.74	$73.38	$102.86
35	$44.28	$71.03	$101.55
40	$41.79	$69.53	$100.85
45	$39.96	$68.56	$100.47

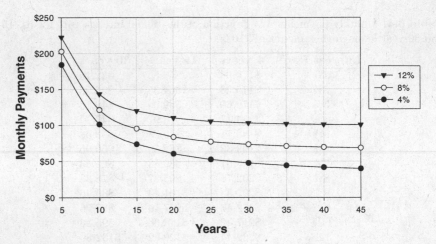

Fig. 6.3. Monthly payments as a function of the length of a $10,000 loan for annual interest rates of 4%, 8%, and 12%

- Because the periodic payment is typically rounded up to the nearest cent, P_N is usually negative, not zero. Therefore, the **final payment**, F, may not be exactly M. Figure 6.4 shows a time diagram for this. Now, F is given by

$$F = P_{N-1}(1+i) = \left(\left(P_0 - \frac{M}{i} \right) (1+i)^{N-1} + \frac{M}{i} \right) (1+i).$$

Thus, the final payment[2] is

$$F = \left(P_0 - \frac{M}{i} \right) (1+i)^N + M\frac{1+i}{i}. \tag{6.6}$$

Intuitively we expect $0 < F \le M$ (why?), but is it? From (6.3) and (6.6) and the fact that $M > iP_0$, we have $0 < F \le M$.

Example 6.1. Amanda Kendrick takes out a student loan for $10,000 at an annual interest rate of 10% to be repaid monthly over 10 years. How much are her monthly payments? What is the final payment?

---

[2] In practice, the final payment may differ slightly from the amount given by F. This could occur if the lender rounds all calculations to two decimal places at every stage. In this case an accumulative round-off error in P_n may come into effect. For example, if $130,000 is borrowed at 8.5% for 180 months and $M = \$1,280.16$, then $F = \$1,280.68$, whereas if all calculations are rounded to two decimal places, then the final payment is $1,280.62. (See Problems 6.4 and 6.5 on p. 98.)

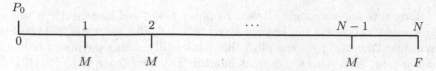

Fig. 6.4. Time diagram of a mortgage

Solution. Here $P_0 = 10,000$, $i = 0.10$ (10%), $m = 12$, and $N = 120$ (10 years), so (6.5) gives

$$M = \frac{\frac{0.10}{12}10000}{1 - \left(1 + \frac{0.10}{12}\right)^{-120}} = 132.151.$$

The amortization table on p. 83 gives \$132.16. It is common practice to round the payments up (why?), so these two numbers agree if we do that. Henceforth, we follow this practice of rounding all amortization payments up.[3]

The final payment is

$$F = \left(10000 - \frac{132.16}{\frac{0.10}{12}}\right)\left(1 + \frac{0.10}{12}\right)^{120} + 132.16\left(\frac{1 + \frac{0.10}{12}}{\frac{0.10}{12}}\right) = \$130.26.$$

The payments total $119 \times \$132.16 + \$130.26 = \$15,857.30$. Thus, Amanda pays \$5,857.30 interest on a \$10,000 loan over 10 years at 10%. △

Example 6.2. Amanda wants to borrow the same \$10,000 at the same annual interest rate of 10%, but she wants to extend her payments over 30 years. How much is her monthly payment?

Solution. Here $P_0 = 10,000$, $i = 0.10$ (10%), $m = 12$, and $N = 360$ (30 years), so (6.5) gives

$$M = \frac{\frac{0.10}{12}10000}{1 - \left(1 + \frac{0.10}{12}\right)^{-360}} = \$87.757.$$

Amortization tables give \$87.76.

The final payment is

$$F = \left(10000 - \frac{87.76}{\frac{0.10}{12}}\right)\left(1 + \frac{0.10}{12}\right)^{360} + 87.76\left(\frac{1 + \frac{0.10}{12}}{\frac{0.10}{12}}\right) = \$81.33.$$

The payments total $359 \times \$87.76 + \$81.34 = \$31,587.18$. Thus, Amanda pays \$21,587.18 interest on a \$10,000 loan over 30 years at 10%. △

---

[3] This is the practice of lending agencies such as the Government National Mortgage Association (GNMA, known as "Ginnie Mae"), the Federal National Mortgage Association (FNMA, known as "Fannie Mae"), and the Federal Home Loan Mortgage Corporation (FHLMC, known as "Freddie Mac"). However, this practice is not universal; see the previous footnote.

Borrowers are commonly advised to repay amortized loans, such as mortgages, every two weeks rather than every month.[4] This results in 26 biweekly payments. One way to accomplish this while still making payments only 12 times a year, is to raise the current monthly payment from M to $(13/12)M$. Let's see what difference that makes to the previous example.

Example 6.3. Under the circumstances of Example 6.2, how many months does it take to repay the loan if Amanda decides to make payments of $(13/12)87.76 = \$95.07$ per month?

Solution. Here $P_0 = 10,000$, $i = 0.10$ (10%), and $M = \$95.07$, so (6.4) gives

$$\frac{\ln\left(\frac{95.07}{95.07 - \frac{0.10}{12}10000}\right)}{\ln\left(1 + \frac{0.10}{12}\right)} \leq N < \frac{\ln\left(\frac{95.07}{95.07 - \frac{0.10}{12}10000}\right)}{\ln\left(1 + \frac{0.10}{12}\right)} + 1,$$

that is,

$$252.07 \leq N < 253.07,$$

so $N = 253$ months (21 years and 1 month). The final payment is

$$F = \left(10000 - \frac{95.07}{\frac{0.10}{12}}\right)\left(1 + \frac{0.10}{12}\right)^{253} + 95.07\left(\frac{1 + \frac{0.10}{12}}{\frac{0.10}{12}}\right) = \$6.87.$$

Thus, the loan is repaid in about 21 years, with payments totaling $252 \times \$95.07 + \$6.87 = \$23,964.51$, which is \$7,500 less than the answer to Example 6.2. $\triangle$

Example 6.4. Helen and Hugh Kendrick pay \$500 a month on a 30-year home loan for \$48,000. What interest rate are they being charged?

Solution. From (6.5), we want to solve

$$500 = \frac{48000i}{1 - (1 + i)^{-360}}$$

for i. There is no simple algebraic way to solve this equation. However, if we write it in the form

$$500\left(1 - (1 + i)^{-360}\right) - 48000i = 0,$$

or

$$(1 + i)^{-360} + 96i - 1 = 0,$$

then we can use Newton's Method[5] or the bisection method to find that $i = 0.0101$, so $i^{(12)} = 12i = 0.121$. (Another solution is $i = 0$. Why is this unacceptable?) $\triangle$

---

[4] In practice, the lending company may impose a penalty for these extra payments.

[5] Newton's Method requires a reasonable initial guess. In the equation $(1 + i)^{-360} + 96i - 1 = 0$ the term $(1 + i)^{-360}$ is negligible compared with the other two terms, so as a first approximation we can assume that $96i - 1 \approx 0$, which leads to $i \approx 1/96$ as an initial guess.

Example 6.5. Helen is in the market for a new car. She sees an advertisement that offers either a \$1,500 discount for paying cash or 0.9% nominal interest on a 60-month amortized loan. Which is the better deal if savings accounts are currently earning 5% per year?

Solution. For Helen to consider the first option, she must have enough cash to buy the car outright. If she chooses the second option, then she could invest her money in a savings account.

We look at this problem in a general setting and calculate how much Helen would have in her savings account if she does not buy the car outright. Let P_0 be the cost of the car, let i be the monthly interest rate on the savings account, and let j the monthly interest on the car loan. Let P be the monthly car payment of which there are N, so from (6.5)

$$P = \frac{jP_0}{1 - (1+j)^{-N}}.$$

At the end of the first month she receives iP_0 interest and pays P, so P_1, the amount in the savings account at the end of the first month, would be

$$P_1 = P_0 + iP_0 - P = (1+i)P_0 - P.$$

At the end of the second month she receives iP_1 interest and pays P, so

$$P_2 = P_1 + iP_1 - P = (1+i)P_1 - P = ((1+i)P_0 - P)(1+i) - P,$$

or

$$P_2 = (1+i)^2 P_0 - P((1+i)+1).$$

Thus, the amount in the savings account at the end of the nth month would be

$$P_n = (1+i)^n P_0 - P\left((1+i)^{n-1} + (1+i)^{n-2} + \cdots + 1\right)$$

$$= (1+i)^n P_0 - \frac{(1+i)^n - 1}{i}P,$$

or

$$P_n = (1+i)^n P_0 - \frac{(1+i)^n - 1}{i}\frac{jP_0}{1-(1+j)^{-N}}$$

$$= (1+i)^n \left(1 - \frac{j}{i}\frac{1-(1+i)^{-n}}{1-(1+j)^{-N}}\right)P_0.$$

For P_N, the amount in the savings account when the car is paid off, to be equivalent to $C(1+i)^N$, the future value of a discount of C dollars, we must have $P_N = C(1+i)^N$, so

$$P_0 = \frac{C}{\left(1 - \frac{j}{i}\frac{1-(1+i)^{-N}}{1-(1+j)^{-N}}\right)}.$$

For $i = 0.05/12$, $j = 0.009/12$, $N = 60$, and $C = 1500$, we have

$$P_0 = \frac{1500}{\left(1 - \frac{0.009}{0.05}\frac{1-\left(1+\frac{0.05}{12}\right)^{-60}}{1-\left(1+\frac{0.009}{12}\right)^{-60}}\right)} = 15548.89.$$

So if the car costs less than \$15,548.89, then Helen should pay cash; otherwise she should choose the loan. $\triangle$

6.3 Linear Interpolation

There are a number of financial tables available that tabulate, in numerical form, some of the formulas we have derived so far. For example, reference [9] contains tables of future values, present values, compound interest comparisons, and amortization tables. Inevitably, the tables cannot contain all possible cases.

What do we do if we are forced to use tables, but the information that we require is not contained in the tables? We approximate. For example, Table 2.3 on p. 16 shows the future value of \$1 compounded annually for different interest rates i and different numbers of years. Part of that is reproduced here as Table 6.6.

Table 6.6. Future Value of \$1

Interest Rate	Years	
	5	15
3%	\$1.159	\$1.558
5%	\$1.276	\$2.079

If all we have is this table and we need the future value of \$1 for 5 years at 4%, which is halfway between 3% and 5%, then we estimate halfway between 1.159 and 1.276, that is, $(1.159 + 1.276)/2 = 1.2175$. The exact number, computed from (2.1) on p. 15, is

$$P_n = P_0(1+i)^n = (1+0.04)^5 = 1.2166\ldots,$$

so the estimate of 1.2175 is reasonable.

On the other hand if we need the future value of \$1 at 3% for 10 years, which is halfway between 5 years and 15 years, then we estimate halfway between 1.159 and 1.558, that is, $(1.159 + 1.558)/2 = 1.3585$. The exact number is

$$P_n = P_0(1+i)^n = (1+0.03)^{10} = 1.3439\ldots,$$

so the estimate of 1.3585 is reasonable.

This estimation technique is called LINEAR INTERPOLATION and can be explained in the following way. Given two points $(a, f(a))$ and $(b, f(b))$ and a number l such that $a < l < b$, how do we estimate $f(l)$? Without any other information, we join the two given points with a line, $y = g(x)$, and then approximate $f(l)$ with $g(l)$. We derive a general formula for $g(l)$ as follows.

The line $y = g(x)$ that joins the points $(a, f(a))$ and $(b, f(b))$ passes through $(a, f(a))$ and has slope $(f(b) - f(a))/(b - a)$; therefore its equation is given by

$$g(x) - f(a) = \frac{f(b) - f(a)}{b - a}(x - a).$$

With $x = l$, where $a < l < b$, we have

$$g(l) = \frac{f(b) - f(a)}{b - a}(l - a) + f(a),$$

which can be rewritten in the form

$$g(l) = \frac{b - l}{b - a}f(a) + \frac{l - a}{b - a}f(b). \tag{6.7}$$

This can be interpreted geometrically if we realize that $(b - l)/(b - a)$ is the ratio of the distance from l to b to the distance from a to b and that $(l - a)/(b - a)$ is the corresponding ratio from a to l. In the event that l is midway between a and b, that is, $l = (a + b)/2$, we find that

$$g\left(\frac{a + b}{2}\right) = \frac{1}{2}f(a) + \frac{1}{2}f(b),$$

the formula we used at the start of this section.

Example 6.6. Use Table 6.6 to estimate the future value of $1 at 3.5% for 15 years. Compare this to the exact number obtained from (2.1) on p. 15.

Solution. Here $a = 3$, $b = 5$, and $l = 3.5$, so

$$g(3.5) = \frac{1.5}{2}f(3) + \frac{0.5}{2}f(5) = \frac{3}{4}1.558 + \frac{1}{4}2.079 = 1.68825.$$

The exact number is

$$P_n = P_0(1 + i)^n = (1 + 0.035)^{15} = 1.6753\ldots.$$

$\triangle$

Again 1.68825 is a reasonable estimate of the exact number 1.6753.... However, from this calculation, we cannot tell how good the estimate is or even whether the estimate is too large or too small. But we notice that all three estimates that we have made are larger than the exact number. Is that an accident? If we look at Fig. 2.1 on p. 17, then we are reminded that the future value is a concave up function of the interest rate. Thus, if we try to estimate a value between two points by joining those two points with a line, then our estimate is too large. A similar observation is true when we think of the future value as a function of the total number of interest periods (see Fig. 2.2 on p. 17).

That $f(l) < g(l)$ in this case is made precise by the following theorem.

Theorem 6.4. *Let $f(x)$ be a twice differentiable function on the closed interval $[a, b]$ with the property that $f''(x) > 0$ for all x in the open interval (a, b). Then for any point l in the open interval (a, b), we have*

$$f(l) < \frac{b - l}{b - a} f(a) + \frac{l - a}{b - a} f(b). \tag{6.8}$$

Proof. By the Mean Value Theorem[6] applied to $f(x)$ in the closed interval $[a, l]$, there must exist a c in (a, l) for which

$$f'(c) = \frac{f(l) - f(a)}{l - a}.$$

Similarly, there must exist a d in (l, b) for which

$$f'(d) = \frac{f(b) - f(l)}{b - l}.$$

Because $f''(x) > 0$, it follows that $f'(x)$ is an increasing function. Also, $a < c < l < d < b$, so $f'(c) < f'(d)$, that is,

$$\frac{f(l) - f(a)}{l - a} < \frac{f(b) - f(l)}{b - l}.$$

Solving this inequality for $f(l)$ shows that

$$f(l) < \frac{b - l}{b - a} f(a) + \frac{l - a}{b - a} f(b).$$

Comments

- The equation of the line $y = g(x)$ that joins the points $(a, f(a))$ and $(b, f(b))$ is

$$g(x) = \frac{b - x}{b - a} f(a) + \frac{x - a}{b - a} f(b),$$

which is the right-hand side of (6.8) when $x = l$. Thus, the right-hand side of (6.8) is the height of the line at $x = l$, whereas the left-hand side, $f(l)$, is the height of the function $f(x)$ at $x = l$. Thus, this theorem proves the intuitively obvious fact that if a positive function is concave up, then the line between the end-points is always higher than the function.

---

[6] Let f be a continuous function on the closed interval $[a, b]$ with a derivative at every x in the open interval (a, b). Then there is at least one number c in the open interval (a, b) such that $f'(c) = (f(b) - f(a))/(b - a)$.

- If we are given the value of the function $f(x)$ at the two points a and b—so we are given the two points $(a, f(a))$ and $(b, f(b))$—and are asked to estimate the value of $f(x)$ at the point $x = l$ between a and b, then a standard way to do this is by linear interpolation, that is, join the two points with a line $y = g(x)$ and use the estimate $g(l)$ for $f(l)$. The theorem tells us that in the case of a positive, concave up function, the estimate $g(l)$ is always too large. In the same way, an estimate obtained by linear interpolation for a positive, concave down function is always too small.
- The inequality (6.8) can be rewritten in the form

$$f(\lambda a + (1-\lambda)b) < \lambda f(a) + (1-\lambda)f(b) \qquad (6.9)$$

for all λ satisfying $0 < \lambda < 1$. (See Problem 6.17.)

6.4 Problems

Walking

6.1. Use a spreadsheet program to reproduce the following amortization table for 10% interest.

Term Amount	10 years	11 years	12 years	13 years	14 years
$10,000	132.16	125.20	119.51	114.79	110.83
$11,000	145.37	137.72	131.46	126.27	121.91

6.2. An amortization schedule is a table with the following (or equivalent) headings.

Month	Monthly Payment	Interest Paid	Principal Repaid	Outstanding Principal

Complete the following partial amortization schedule.

Month	Monthly Payment	Interest Paid	Principal Repaid	Outstanding Principal
1	$188.12	$55.42	$132.70	$9,367.30
2				
3				
4				
5				

6.3. You buy a house by borrowing $100,000 at 8% over 30 years. Show the first two and last two lines of the amortization schedule. Is the final payment the same as the monthly payment?

6.4. Use a spreadsheet program to construct an amortization schedule for a 15-year loan for $130,000 at 8.5% with monthly payments of $1,280.16. What is the final payment? Does it agree with (6.6) on p. 90? Why does the final payment exceed the monthly payment? What monthly payment guarantees that the final payment is less than the monthly payment?

6.5. Repeat Problem 6.4, but now round the entries in the *Interest Paid* column to 2 decimal places. What is the final payment? Does it agree with (6.6) on p. 90?

6.6. A person pays $200.04 a month on a 15-year loan for $30,000. What interest rate is being charged?

6.7. In [10], a 30-year loan for $75,000 at 10% is discussed. On page 51 it is claimed that paying $25 a month more than the regular monthly amount reduces the term of the mortgage by 12 years and 3 months, while saving $34,162 in interest costs. Is this claim valid?

6.8. Use linear interpolation to estimate the monthly loan payment on $45,000 at 10% for 30 years given that the monthly loan payment on $100,000 at 10% for 30 years is $877.58. Is your estimate exact, too large, or too small?

6.9. Use linear interpolation to complete the following amortization table. Indicate which of your entries are exact, too large, or too small

(a) Without calculating exact values.
(b) By calculating exact values.

Term Amount	10 years	11 years	12 years	13 years	14 years
$1,000					
$10,000	$132.16				$110.83
$11,000					

Given that $m=12$, that is, payments are made monthly, what is the annual interest rate?

Running

6.10. Intuitively you might expect that if you double the amount of money that you borrow (at the same rate over the same number of years), then you have to double your monthly payments. Prove that this is true.

6.11. Intuitively you might expect that if you double the monthly payments, then your loan is repaid in half the time. Is this true?

6.12. Prove (6.2) on p. 86 by induction.

6.13. Show that if we finish repaying a loan exactly at the end of the Nth period, so that $P_N = 0$, then (6.2) can be written in the form

$$P_n = \frac{M}{i}\left(1 - (1+i)^{n-N}\right).$$

6.14. Consider the following eight loans, with initial principals (amounts borrowed) of $10,000 and $20,000, maturities of 10 and 25 years, and interest rates of 10% and 7.5%.

Amount Borrowed	Maturity	Interest Rate	Monthly Payment
$10,000	10	10.0%	
$10,000	10	7.5%	
$10,000	25	10.0%	
$10,000	25	7.5%	
$20,000	10	10.0%	
$20,000	10	7.5%	
$20,000	25	10.0%	$181.75
$20,000	25	7.5%	

The monthly payment for the $20,000 loan with a maturity of 25 years and an interest rate of 10% is $181.75. The remaining monthly payments are

$$\$73.90, \$90.88, \$118.71, \$132.16, \$147.80, \$237.41, \text{ and } \$264.31,$$

although not necessarily in this order. Without using a calculator, complete the table.

6.15. Show that by substituting $i = x - 1$ and $N = n + 1$ into (6.5) on p. 88, we have

$$\frac{M}{P_0} = \frac{i}{1 - (1 + i)^{-N}} = \frac{(x - 1)x^{n+1}}{x^{n+1} - 1} = \frac{x^{n+1}}{\sum_{k=0}^{n} x^k}.$$

(a) Show that

$$\frac{d}{dx}\left(\frac{M}{P_0}\right) = \frac{1}{(\sum_{k=0}^{n} x^k)^2} \sum_{k=0}^{n}(n + 1 - k)x^{n+k}.$$

Explain how this shows that M is an increasing function of i if $i > 0$.

(b) Show that

$$\frac{d^2}{dx^2}\left(\frac{M}{P_0}\right) = \frac{1}{(\sum_{k=0}^{n} x^k)^3} \sum_{k=0}^{n}\sum_{h=0}^{n}(n + 1 - k)(n + k - 2h)x^{n+k+h-1}.$$

(c) Show that

$$\sum_{k=0}^{n}\sum_{h=0}^{n} a_{kh}x^{n+k+h-1} = \sum_{j=0}^{2n} b_j x^{n+j-1},$$

where $b_j = \sum_{k=0}^{j} a_{kj-k}$ if $0 \le j \le n$ and $b_j = \sum_{k=0}^{2n-j} a_{n-k,j+k-n}$ if $n \le j \le 2n$.

(d) Use parts (b) and (c) to show that $b_j = (j+1)(n+1)(n-j)$ if $0 \leq j \leq n$ and $b_j = 0$ if $n \leq j \leq 2n$.

(e) Explain how parts (b), (c), and (d) show that M is a concave up function of i if $i > 0$.

6.16. Show that by substituting $N = x$ and $1 + i = a$ into (6.5) on p. 88, we have

$$\frac{M}{iP_0} = \frac{1}{1 - a^{-x}} = 1 + \frac{1}{a^x - 1}.$$

(a) Show that

$$\frac{d}{dx}\left(\frac{M}{iP_0}\right) = -\frac{a^x \ln a}{(a^x - 1)^2}.$$

Explain how this shows that M is a decreasing function of the positive integer N.

(b) Show that

$$\frac{d^2}{dx^2}\left(\frac{M}{iP_0}\right) = \frac{(\ln a)^2 (a^{2x} + a^x)}{(a^x - 1)^3}.$$

Explain how this shows that M is a concave up function of i if $i > 0$.

6.17. By introducing $\lambda = (b - l)/(b - a)$, show that $1 - \lambda = (l - a)/(b - a)$ and that $l = \lambda a + (1 - \lambda) b$. Now show that (6.9) is a consequence of (6.8).

6.18. Show that for a concave down function $f(x)$ if we estimate the value of $f(x)$ at a point x with $a < x < b$ using linear interpolation, then the estimate is too small.

Questions for Review

- What is amortization?
- What is an amortization table?
- What is an amortization schedule?
- How do you calculate the principal remaining to be paid after n periods?
- What is the impact on the outstanding principal if $M > iP_0$? If $M = iP_0$? If $M < iP_0$?
- How do you calculate the number of months required to repay the loan?
- How do you calculate the periodic payment on a loan?
- How do you calculate the final payment on a loan?
- What does the Amortization Theorem say?
- What is linear interpolation?
- What is the Mean Value Theorem?
- How do you use linear interpolation to estimate the value of a function at a point c in (a, b)?
- Why are the estimates of the future value of $1 obtained by linear interpolation always too high?

7

Credit Cards

A credit card account is an unsecured revolving line of credit issued by a financial institution to an entity.[1] Credit cards are often used for purchases of goods and services when payment by cash is not convenient or permissible. Examples of such transactions include on-line or telephone purchases of goods and services, business expenditures, reservations for hotel rooms, and car rentals.

The loan is unsecured, so there is no collateral that the bank can seize if the borrower defaults on the loan. The loan is also revolving, which means that the borrower can continue to borrow against the line of credit as long as there are available funds. These aspects cause credit cards to be riskier from the financial institution's perspective than mortgages, car loans, and other secured installment loans. Therefore, the interest rates charged on credit cards tend to be higher than those charged on secured loans. In return for the borrowed funds, the borrower promises to pay each month either a fixed amount or a percentage of the outstanding balance, whichever is higher. In general, credit card disclosures contain a statement similar to "the minimum payment will be p or $r\%$ of the balance remaining to be paid, whichever is higher".

In this chapter we show how credit card payments are computed.

7.1 Credit Card Payments

We charge $1,000 to a credit card with an annual interest rate of 15% for which the minimum monthly payment is $10 or 2% of the balance remaining to be paid, whichever is higher. We do not charge anything more to the card and decide to make only the minimum payment each month.

---

[1] An UNSECURED LOAN is a loan that is not backed by collateral, a mortgage, or other lien. A LINE OF CREDIT is an agreement between a lender and a borrower establishing a maximum balance that the lender permits the borrower to carry.

When the first month's bill arrives we pay the higher of \$10 and \$1,000 × 0.02 = \$20, namely, \$20. The outstanding balance when next month's bill arrives is \$1,000 − \$20 = \$980. We pay the higher of \$10 and \$980 × 0.02 = \$19.60, namely, \$19.60. However, the outstanding balance is \$980, so we owe \$980 × 0.15/12 = \$12.25 in interest. Thus, the outstanding balance in the third month is \$980 − \$19.60 + \$12.25 = \$972.65. With this information, we complete Table 7.1, which shows the first five payments. (Check the calculations in this table using a calculator or a spreadsheet program, and fill in the missing entries.)

Table 7.1. Credit Card Schedule

Month	Monthly Payment	Interest Accrued	Remaining Principal
0	\$0.00	\$0.00	\$1,000.00
1	\$20.00	\$0.00	\$980.00
2	\$19.60	\$12.25	\$972.65
3	\$19.45	\$12.16	\$965.36
4			
5	\$19.16	\$11.98	\$950.94

Based on this, we let

- P_0 be the INITIAL PRINCIPAL (initial balance) that has to be repaid,
- P_n be the PRINCIPAL REMAINING to be paid (outstanding balance) at the end of the nth month,
- $i^{(12)}$ be the fixed ANNUAL INTEREST RATE, expressed as a decimal,
- i be the MONTHLY INTEREST RATE, $\left(i = i^{(12)}/12\right)$,
- p be the MINIMUM MONTHLY PAYMENT, and
- r be the MINIMUM MONTHLY BALANCE REPAYMENT RATE, expressed as a decimal.

Using this notation we rewrite Table 7.1 symbolically in spreadsheet format as Table 7.2, which is explained as follows.

Table 7.2. Credit Card Schedule—Spreadsheet Format

Month	Monthly Payment	Interest Accrued	Remaining Principal
0	0	0	P_0
1	rP_0	0	$P_0 - rP_0 = P_1$
2	rP_1	iP_1	$P_1 - rP_1 + iP_1 = P_2$
3	rP_2	iP_2	$P_2 - rP_2 + iP_2 = P_3$
4	rP_3	iP_3	$P_3 - rP_3 + iP_3 = P_4$
5	rP_4	iP_4	$P_4 - rP_4 + iP_4 = P_5$

We concentrate on the case where we charge P_0 to this credit card and then charge nothing else but elect to pay the minimum each month. Also, initially we assume that $rP_0 > p$. (See Problem 7.8 on p. 111 for the case in which $rP_0 \le p$.)

At the end of the first month ($n = 1$) we make our monthly payment of rP_0 (assuming that $rP_0 > p$), and because there is no interest, the principal remaining to be paid is

$$P_1 = P_0 - rP_0 = P_0(1 - r).$$

At the end of the second month ($n = 2$) we make our monthly payment of rP_1 (assuming that $rP_1 > p$), but the interest that we owe on the outstanding balance P_1 is iP_1, so the principal remaining to be paid is

$$P_2 = P_1 - rP_1 + iP_1 = P_1(1 - r + i) = P_0(1 - r + i)(1 - r).$$

At the end of the third month ($n = 3$) the interest that we owe is iP_2, so the principal remaining to be paid is

$$P_3 = P_2 - rP_2 + iP_2,$$

which reduces to

$$P_3 = P_0(1 - r + i)^2(1 - r).$$

This suggests that P_n, the principal remaining to be paid at the end of the nth month, is

$$P_n = P_0(1 - r + i)^{n-1}(1 - r). \tag{7.1}$$

We make three observations about this formula.

1. If $i > r$, then $P_{n+1} > P_n$ and $\lim_{n \to \infty} P_n = \infty$, that is, $\{P_n\}$ is an increasing sequence that is unbounded. In other words, the amount owed increases each month. Not only does it increase, it increases at an exponential rate.
2. If $i = r$, then $P_n = P_0(1 - r)$ for all n, so the amount owed remains constant.
3. If $i < r$, then $P_{n+1} < P_n$, $P_n > 0$, and $\lim_{n \to \infty} P_n = 0$, that is, it takes an infinite time to repay the loan. We note that $rP_{n+1} < rP_n$, so our payments always decrease. However, we see that the amount going to pay off the debt, namely $(r - i)P_n$, is also decreasing. Finally, the proportion i of our payment going to interest every month is constant.

However, credit cards have a statement to the effect that the minimum monthly payment is either p or $r\%$ of the balance remaining to be paid, whichever is higher. Thus, in the third case there is a month N_1 when, for the last time, our proposed monthly payment (rP_{N_1-1}) is greater than p, that is, $rP_{N_1-1} > p$. Because N_1 is the last month in which this occurs, the next monthly payment (rP_{N_1}) cannot be greater than p, that is, $rP_{N_1} \leq p$. Thus, the integer N_1 must satisfy

$$rP_{N_1} \leq p < rP_{N_1-1},$$

or

$$P_0 \left(1 - r + i\right)^{N_1-1} \left(1 - r\right) \leq \frac{p}{r} < P_0 \left(1 - r + i\right)^{N_1-2} \left(1 - r\right).$$

Because $\ln\left(1 - r + i\right) < 0$, we have

$$N_1 - 2 < \frac{\ln\left(\frac{p}{P_0 r(1-r)}\right)}{\ln\left(1 - r + i\right)} \leq N_1 - 1,$$

or N_1, the month when we switch from decreasing payments to the constant payment of p, is determined from

$$1 + \frac{\ln\left(\frac{p}{P_0 r(1-r)}\right)}{\ln\left(1 - r + i\right)} \leq N_1 < 2 + \frac{\ln\left(\frac{p}{P_0 r(1-r)}\right)}{\ln\left(1 - r + i\right)}. \tag{7.2}$$

Because we have assumed that $rP_1 > p$, we have $p/(P_0 r \left(1 - r\right)) < 1$, which requires that $\ln\left(p/(P_0 r \left(1 - r\right))\right) < 0$ (why?), so that

$$\frac{\ln\left(\frac{p}{P_0 r(1-r)}\right)}{\ln\left(1 - r + i\right)} > 0.$$

Thus, (7.2) has an integer solution $N_1 \geq 2$. (See Problems 7.9 and 7.8 on p. 111 for the case in which $rP_1 \leq p$.)

In this case

$$P_{N_1+1} = P_{N_1} - p + iP_{N_1} = P_{N_1}(1 + i) - p.$$

However, because $i < r$, we then have

$$P_{N_1+1} = P_{N_1}(1 + i) - p \leq (1 + i)\frac{p}{r} - p = \frac{p}{r}(1 + i - r) < \frac{p}{r},$$

so $P_{N_1+1} < p/r$. Thus, once the constant payment starts it remains in effect, giving

$$P_{N_1+2} = P_{N_1+1} - p + iP_{N_1+1} = P_{N_1+1}(1 + i) - p,$$

which can be written as

$$P_{N_1+2} = (P_{N_1}(1 + i) - p)(1 + i) - p = P_{N_1}(1 + i)^2 - p(1 + (1 + i)).$$

Similarly,

$$P_{N_1+3} = P_{N_1+2} - p + iP_{N_1+2} = P_{N_1}(1 + i)^3 - p(1 + (1 + i) + (1 + i)^2).$$

Thus, we expect, once $P_{N_1} \le p/r$, that

$$P_{N_1+n} = P_{N_1}(1+i)^n - p\left(1 + (1+i) + \cdots + (1+i)^{n-1}\right),$$

or, using the fact that for $x \ne 1$,

$$1 + x + \cdots + x^{n-1} = \frac{x^n - 1}{x - 1},$$

we have

$$P_{N_1+n} = P_{N_1}(1+i)^n - p\frac{(1+i)^n - 1}{i}. \tag{7.3}$$

Except for notation, this is exactly the same equation as (6.1) on p. 85 and is proved in the same way, leading to a result similar to (6.4). We thus have the following results.

Theorem 7.1. *The Credit Card Theorem.*
Assume that $rP_0 > p$ and $rP_1 > p$.

(a) If $i > r$, then

$$P_n = P_0 (1 - r + i)^{n-1} (1 - r)$$

and $P_{n+1} > P_n$, so the amount owed increases each month.
(b) If $i = r$, then

$$P_n = P_0 (1 - r)$$

and $P_{n+1} = P_n$, so the amount owed remains constant each month.
(c) If $i < r$, then we have

$$P_n = \begin{cases} P_0 (1 - r + i)^{n-1} (1 - r) & \text{for } 1 \le n \le N_1, \\ \left(P_{N_1} - \frac{p}{i}\right)(1+i)^{n-N_1} + \frac{p}{i} & \text{for } N_1 < n, \end{cases}$$

where the month N_1 is determined by (7.2). In this case the amount charged is paid off in $N_1 + N_2$ months, where

$$\frac{\ln\left(\frac{p}{p - iP_{N_1}}\right)}{\ln(1+i)} \le N_2 < \frac{\ln\left(\frac{p}{p - iP_{N_1}}\right)}{\ln(1+i)} + 1. \tag{7.4}$$

Comments About the Credit Card Theorem

- If part (c) of the theorem applies, then the total amount paid is approximately[2] (see Problem 7.13 on p. 111)

$$rP_0 \left(1 + \frac{(1 - r)\left(1 - (1 - r + i)^{N_1-1}\right)}{r \quad i}\right) + N_2 p. \tag{7.5}$$

---

[2] This formula assumes that the final payment is exactly p. However, as with amortization, the final payment may be smaller than p.

- The Credit Card Theorem is also valid if the borrower decides to pay a larger p, the minimum monthly payment, and a larger r, the minimum monthly balance repayment rate, than the credit card company requires.

Example 7.1. Tom charges \$1,000 on a credit card with an annual interest rate of 15% and for which the minimum payment is \$10 or 2% of the balance remaining to be paid, whichever is higher. If Tom charges no more on this card but makes the minimum payment each month, then how long does it take for him to repay the debt, and approximately how much does he pay in total?

Solution. Here $P_0 = 1,000$, $i^{(12)} = 0.15$ (15%), $r = 0.02$ (2%), and $p = 10$. First, we see that $rP_0 = 20$ is greater than p and that $rP_1 = P_0 r (1 - r) = \$19.60$ is greater than p, so the Credit Card Theorem applies. Second, we see that $i = 0.15/12 = 0.0125$ is less than 0.02, so part (c) of that theorem applies. We find N_1 from (7.2),

$$1 + \frac{\ln \left(\frac{10}{1000 \times 0.02(1-0.02)} \right)}{\ln \left(1 - 0.02 + \frac{0.15}{12} \right)} \le N_1 < 2 + \frac{\ln \left(\frac{10}{1000 \times 0.02(1-0.02)} \right)}{\ln \left(1 - 0.02 + \frac{0.15}{12} \right)}.$$

Now,

$$\frac{\ln \left(\frac{10}{1000 \times 0.02(1-0.02)} \right)}{\ln \left(1 - 0.02 + \frac{0.15}{12} \right)} = 89.389,$$

so

$$90.389 \le N_1 < 91.389,$$

giving $N_1 = 91$. It takes 91 months (7 years 7 months) to switch from decreasing minimum payments to constant ones. We find N_2 from (7.4),

$$\frac{\ln \left(\frac{10}{10 - \frac{0.15}{12} P_{91}} \right)}{\ln \left(1 + \frac{0.15}{12} \right)} \le N_2 < \frac{\ln \left(\frac{10}{10 - \frac{0.15}{12} P_{91}} \right)}{\ln \left(1 + \frac{0.15}{12} \right)} + 1,$$

where from (7.1),

$$P_{91} = 1000 \left(1 - 0.02 + \frac{0.15}{12} \right)^{91-1} (1 - 0.02) = 497.71,$$

so

$$\frac{\ln \left(\frac{10}{10 - \frac{0.15}{12} P_{91}} \right)}{\ln \left(1 + \frac{0.15}{12} \right)} = 78.34.$$

Thus,

$$78.34 \le N_2 < 79.34,$$

so $N_2 = 79$. Thus, $N_1 + N_2 = 91 + 79 = 170$, so it takes 170 months (14 years and 2 months) to repay the loan.

The total amount paid is given approximately[3] by (7.5), namely,

$$0.02(1000)\left(1+\frac{(1-0.02)\left(1-(1-0.02+0.0125)^{90}\right)}{0.02-0.0125}\right)+79(10)=2096.119.$$

So on the $1,000 loan, Tom pays about $2,096.12. △

We now repeat Example 7.1, to see what happens if Tom decides to pay more than the minimum each month.

Example 7.2. Tom charges $1,000 on a credit card with an annual interest rate of 15% and for which the minimum payment is $10 or 2% of the balance remaining to be paid, whichever is higher. If Tom charges no more on this card but pays twice the minimum payment each month, then how long does it take Tom to repay the debt and approximately how much does he pay in total?

Solution. Here $P_0 = 1000$, $i^{(12)} = 0.15$ (15%), $r = 0.04$ (twice 2%), and $p = 20$ (twice 10). First, we see that $rP_0 = 40$ is greater than p, and that $rP_1 = P_0r(1-r) = \$38.40$ is greater than p, so the Credit Card Theorem applies. Second, we see that $i = 0.15/12 = 0.0125$ is less than 0.04, so part (c) of that theorem applies.

We find N_1 from (7.2),

$$1+\frac{\ln\left(\frac{20}{1000\times0.04(1-0.04)}\right)}{\ln\left(1-0.04+\frac{0.15}{12}\right)}\leq N_1<2+\frac{\ln\left(\frac{20}{1000\times0.04(1-0.04)}\right)}{\ln\left(1-0.04+\frac{0.15}{12}\right)}.$$

Now,

$$\frac{\ln\left(\frac{20}{1000\times0.04(1-0.04)}\right)}{\ln\left(1-0.04+\frac{0.15}{12}\right)}=23.39,$$

so

$$24.39\leq N_1<25.39,$$

giving $N_1 = 25$. It takes 25 months (2 years 1 month) to switch from decreasing minimum payments to constant ones.

We find N_2 from (7.4),

$$\frac{\ln\left(\frac{20}{20-\frac{0.15}{12}P_{25}}\right)}{\ln\left(1+\frac{0.15}{12}\right)}\leq N_2<\frac{\ln\left(\frac{20}{20-\frac{0.15}{12}P_{25}}\right)}{\ln\left(1+\frac{0.15}{12}\right)}+1,$$

---

[3] In fact, the final payment is not $10, but $P_{169} = \$3.40$. (See Problem 7.1 on p. 110.)

where from (7.1),

$$P_{25} = 1000 \left(1 - 0.04 + \frac{0.15}{12}\right)^{25-1} (1 - 0.04) = 491.61,$$

so

$$\frac{\ln\left(\frac{20}{20 - \frac{0.15}{12} P_{25}}\right)}{\ln\left(1 + \frac{0.15}{12}\right)} = 29.55.$$

Thus,

$$29.55 \leq N_2 < 30.55,$$

so $N_2 = 30$. Thus, $N_1 + N_2 = 25 + 30 = 55$, so it takes 55 months (4 years and 7 months) to repay the loan.

The total amount paid is given approximately by (7.5), namely,

$$0.04(1000)\left(1 + \frac{(1 - 0.04)\left(1 - (1 - 0.04 + 0.0125)^{24}\right)}{0.04 - 0.0125}\right) + 30(20) = 1321.293.$$

So on the $1,000 loan Tom pays about $1,321.29.

Thus, by doubling the payments, the time to repay the loan drops from 14 years and 2 months to 4 years and 7 months and the amount paid drops from $2,096.12 to $1,321.29. △

7.2 Credit Card Numbers

The digits on a credit card are not chosen randomly, and many of the popular ones (VISA, Mastercard, American Express, Diners Club, Discover) follow a similar pattern. The total number of digits (usually between 13 and 16) is specified by the company. The initial digits determine the type of card, the next digits the issuing financial institution, and the remaining digits (except for the final one) identifies the customer. The final digit is a checksum. It is computed from the previous digits in such a way so as to reduce the risk of incorrectly entering a digit, transposing adjacent digits, or simply entering an invalid card number.

The technique that is used to compute the checksum is called the LUHN ALGORITHM or the Mod 10 Method.[4] To demonstrate the algorithm, consider the credit card number 9876543219876543, so that 3 is the checksum. We concentrate on the other 15 digits. Starting from the right-hand side we multiply every other digit by 2 and leave the remaining digits unchanged, giving

$$
\begin{array}{cccccccc}
9\,8 & 7\,6 & 5\,4 & 3\,2 & 1\,9 & 8\,7 & 6\,5 & 4 \\
\times 2 & \times 2 & \times 2 & \times 2 & \times 2 & \times 2 & \times 2 & \times 2. \\
18\,8 & 14\,6 & 10\,4 & 6\,2 & 2\,9 & 16\,7 & 12\,5 & 8
\end{array}
$$

---

[4] Canadian Social Insurance Numbers, which are nine digits long, also use the Luhn algorithm to check the last digit.

Now, we add the resulting digits, treating each digit separately, so that 18 is treated as $1 + 8$. Thus,

$$(1 + 8)+8+(1 + 4)+6+(1 + 0)+4+6+2+2+9+(1 + 6)+7+(1 + 2)+5+8 = 82.$$

We now add the checksum, 3, to this total to find 85. If the new total is divisible by 10, then the credit card number has passed the validation test.[5] This number fails, whereas 9876543219876548 passes.

In order to analyze whether this algorithm detects the incorrect entry of a digit or the transposition of adjacent digits, we look at what each of the numbers from 0 to 9 transform into as a result of the "multiply by two and add the digits" rule.

Initial Digit d	$\times 2$ $2d$	Final Digit $f(d)$
0	0	0
1	2	2
2	4	4
3	6	6
4	8	8
5	10	1
6	12	3
7	14	5
8	16	7
9	18	9

Imagine that we inadvertently change one digit in an otherwise valid credit card number. If it is the check digit that is changed, then the new number is no longer valid because the sum of the remaining digits requires the correct check digit to pass the test. If it is one of the undoubled digits that is changed, then the new sum differs from the old one by a number less than 10, so it does not give the correct check digit. The same is true if one of the doubled digits is changed.

Imagine that we inadvertently transpose adjacent digits in an otherwise valid credit card number. There are 100 possible combinations of two digits, from 00 through 99, of which ten (00, 11, 22, ..., 99) won't cause any problems if the digits are transposed. If d_1 and d_2 are adjacent digits in the credit card number ($d_1 \neq d_2$), then their contribution to the checksum is either $f(d_1)+d_2$ or $f(d_2) + d_1$, and if transposed they are either $f(d_2) + d_1$ or $f(d_1) + d_2$. If these two sums are the same or their difference is a multiple of 10, then the checksum does not distinguish between their transposition. This suggests that we should look at $(f(d_1) + d_2) - (f(d_2) + d_1) = (f(d_1) - d_1) - (f(d_2) - d_2)$. In other words, we should consider $f(d) - d$.

---

[5] Of course, this does not check whether the credit card is valid.

d	$f(d)$	$f(d) - d$
0	0	0
1	2	1
2	4	2
3	6	3
4	8	4
5	1	-4
6	3	-3
7	5	-2
8	7	-1
9	9	0

One question that we can ask is, "Are there two different d's, called d_1 and d_2, for which $(f(d_1) - d_1) - (f(d_2) - d_2) = 0$?" Looking at the preceding table, we can answer "Yes, 0 and 9." Thus, if the adjacent digits are 0 and 9 and they are interchanged, then the checksum does not pick up this error. Another question we can ask is, "Are there two different d's, called d_1 and d_2, for which $(f(d_1) - d_1) - (f(d_2) - d_2) \neq 0$ but is divisible by 10?" Looking at the preceding table, we can answer "No." Thus, the Luhn algorithm detects the transposition of adjacent digits if they are not 09 or 90.

7.3 Problems

Walking

7.1. Show that the final payment in Example 7.1 on p. 106 is P_{169} and that it is \$3.40.

7.2. We charge \$2,100 on a credit card with an annual interest rate of 18% and for which the minimum payment is \$10 or 2.5% of the balance remaining to be paid, whichever is higher. How long will it take to repay the debt assuming that the minimum payment is always made, and how much will we have paid in total? If we borrow the same \$2,100 at the same annual interest rate of 18% but repay it as an amortized loan with a monthly payment of \$55, how long will it take to repay the debt, and how much will we have paid in total?

7.3. Using Table 7.2 on p. 102 as a template, construct a spreadsheet that reproduces the calculations in Example 7.1 on p. 106 and in Problem 7.2.

7.4. Use (7.5) on p. 105 and the Credit Card Theorem on p. 105 to construct a spreadsheet that calculates N_1, N_2, and the (approximate) total amount repaid after the user inputs P_0, $i^{(12)}$, p, and r. Use the format in the following table.

Amount owed =	$P_0 =$	$1,000.00
Annual interest rate =	$i^{(12)} =$	15%
Minimum monthly payment =	$p =$	$10.00
Minimum monthly balance repayment rate =	$r =$	2%
$i < r$?		Yes
$rP_0 > p$?		Yes
$rP_1 > p$?		Yes
	N_1	91
	P_{N_1}	$491.71
	N_2	79
Months to repay =	$N_1 + N_2$	170
Total repaid =		$2,096.12

7.5. Apply the Luhn algorithm to 9876543219876453. Is the checksum a valid one? If not, then what should the final digit be?

7.6. If 9876543219076545 has a valid checksum using the Luhn Algorithm, will 9876543210976545?

7.7. Apply the Luhn algorithm to 0000000000000000. Is the checksum a valid one?

Running

7.8. State and prove a result similar to the Credit Card Theorem on p. 105 if initially $rP_0 \le p$.

7.9. State and prove a result similar to the Credit Card Theorem on p. 105 if initially $rP_0 > p$ but $rP_1 \le p$.

7.10. Determine what happens if instead of a single charge of P_0 to a credit card, we charge P_0 every month but still elect to pay the minimum each month.

7.11. Determine what happens if in addition to paying the minimum amount each month, we add a fixed dollar amount M.

7.12. Sometimes a credit card company lets a borrower skip a monthly payment (but not the interest). How does this affect the total interest paid?

7.13. Assuming that the final payment is exactly p, prove (7.5) on p. 105.

7.14. Sometimes in the Luhn algorithm, the "multiply by two and add the digits" rule, is replaced by "multiply by two and, if the resulting number is greater than 9, subtract 9" rule. Are these rules the same? Verify.

7.15. Sometimes in the Luhn algorithm, the "multiply by two and add the digits" rule, is replaced by "multiply by two, divide the resulting number by 9, and keep the remainder" rule. Are these rules the same? Verify.

Questions for Review

- How do you find the total amount paid when $i < r$?
- What does the Credit Card Theorem say?
- What is the Luhn algorithm?
- What is the purpose of the Luhn algorithm?
- How does the Luhn algorithm work?
- What is a checksum?
- How is the outstanding balance on a credit card computed?
- How are the payments on the outstanding balance on a credit card computed?
- How does the remaining balance on a credit card change when $i > r$? When $i = r$? When $i < r$?
- How do you find the month in which you switch from decreasing payments to constant payments?
- How do you find the number of months in which the minimum monthly payment, p, is made?

8

Bonds

Bonds are loans that an investor makes either to a government agency (US Treasury bonds or notes[1] or Municipal bonds) or to a corporation (Corporate bonds). The borrower promises to pay the lender interest at regular intervals (usually every six months) and to repay the principal at the end of the loan. In investor's language the previous sentence reads, "The borrower promises to pay the lender the **coupons** on the **coupon payment dates** and to pay the **face value** of the bond at the **redemption date**."

Government Bonds and Notes	
Typical Term	2 years, 3 years, 5 years, 10 years, 30 years
Payment Frequency	Semi-annually
Penalty	None
Issuer	Federal Government
Risks	Inflation, Interest Rate, Reinvestment, Maturity
Marketable	Yes
Restrictions	Minimum Investment

Corporate Bonds	
Typical Term	10 to 30 years
Payment Frequency	Semi-annually
Penalty	None
Issuer	Corporations
Risks	Inflation, Market, Interest Rate, Liquidity, Default, Business, Reinvestment, Maturity
Marketable	Yes
Restrictions	Minimum Investment

---

[1] US Treasury bonds are obligations of the United States government with a maturity of more than 10 years. US Treasury notes are obligations of the United States government with a maturity of between 1 and 10 years.

There are various types of bonds, but one which is not redeemable before the scheduled redemption date is called NONCALLABLE. In this book we concentrate on noncallable bonds.

8.1 Noncallable Bonds

The value of a noncallable bond varies with time. When a bond is first issued, the interest rate is determined by the competitive interest rates of other bonds available at that time. This determines the initial value of the bond—the **primary market value**. Later this bond may be offered for resale. Then its resale value—the **secondary market value**—is determined by the competitive interest rates available at that time, as well as the proportion of the coupon period that has elapsed by the RECORD DATE.[2]

For example, suppose that an investor purchases a 20-year bond for $1,000 when the interest rate is 7%. Later the competitive interest rate falls to 6.5%, and the investor decides to sell the bond. Clearly the investor expects more than $1,000 for the bond because bonds with coupon rates of 6.5% are selling for $1,000. However, if the interest rate rises to 7.5%, then no one will pay $1,000 for the investor's bond because it is worth less than $1,000. But what is a fair price for these bonds? In other words, how do we value a bond that is for sale? This section is devoted to answering that question.

When a bond is sold its value is reported as a percent of its face value; for example, the price of a $1,000 bond may be reported as 98, which means that the market value of the bond is $98/100 \times \$1,000 = \$980$. A price of 100 means that the value of the bond is its face value.

Let

F be the FACE VALUE (par value, redemption value) of the bond,

P be the PRICE of the bond (percentage of the face value),

n be the total NUMBER of coupon payment periods,

m be the number of COUPON PAYMENTS PER YEAR,

$r^{(m)}$ be the ANNUAL COUPON RATE (nominal yield), expressed as a decimal,

r be the COUPON RATE PER COUPON PAYMENT PERIOD,
      expressed as a decimal,

$y^{(m)}$ be the ANNUAL YIELD TO MATURITY (investor's rate of return),
      expressed as a decimal, and

y be the YIELD TO MATURITY PER COUPON PAYMENT PERIOD,
      expressed as a decimal.

---

[2] The Record Date is the date on which a purchaser of a bond becomes the legal owner.

Notice that P is expressed as a percentage, whereas r and y are expressed as decimals,[3] and that $y = y^{(m)}/m$ and $r = r^{(m)}/m$. The cost of the bond is $(P/100)F$, and the coupon payment is rF.

Initially we assume that the price, P, is to be determined immediately following a coupon payment, so the next coupon payment is due exactly one payment period in the future and there are n coupon payment periods remaining. Later we take into account what happens to the price if the bond is sold between coupon payment periods.

The future value of the first coupon payment, rF, on the maturity date is $rF(1+y)^{n-1}$. The future value of the second coupon payment, rF, on the maturity date is $rF(1+y)^{n-2}$. The future value of the final coupon payment, rF, on the maturity date is rF. The future value of the face value, F, is F. The future value of the cost of the bond, $(P/100)F$, on the maturity date is $(P/100)F(1+y)^n$. This is represented by the time diagram shown in Fig. 8.1.

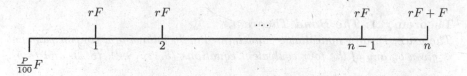

Fig. 8.1. Time diagram of a typical bond

Thus,

$$\frac{P}{100}F(1+y)^n = rF\left((1+y)^{n-1} + (1+y)^{n-2} + \cdots + (1+y) + 1\right) + F,$$

or

$$\frac{P}{100}(1+y)^n = r\left((1+y)^{n-1} + (1+y)^{n-2} + \cdots + (1+y) + 1\right) + 1. \quad (8.1)$$

This equation can be rewritten in various ways. If we multiply by $100/(1+y)^n$, then we have

$$P = 100r\left(\frac{1}{1+y} + \frac{1}{(1+y)^2} + \cdots + \frac{1}{(1+y)^n}\right) + \frac{100}{(1+y)^n}, \quad (8.2)$$

or

$$P = 100r\sum_{k=1}^{n}(1+y)^{-k} + 100(1+y)^{-n}, \quad (8.3)$$

---

[3] Sometimes, instead of being measured in decimal or percentage, rates and yields are measured in **basis points**. A basis point is 0.01%, that is, 0.0001. Basis points are frequently used by dealers dealing with billions of dollars of bonds, and a change of one basis point on one billion dollars is $100,000.

which can be interpreted as "the price is the present value of the expected future cash flows". By rewriting (8.2) in the form

$$P = \frac{100r}{1+y} \left(1 + \frac{1}{1+y} + \cdots + \frac{1}{(1+y)^{n-1}} \right) + \frac{100}{(1+y)^n}$$

and recognizing the expression in parentheses as a geometric series, we find, after a little algebra, that for $y \neq 0$,

$$P = 100 \left(r \frac{1-(1+y)^{-n}}{y} + (1+y)^{-n} \right), \tag{8.4}$$

or

$$P = 100 \left((r-y) \frac{1-(1+y)^{-n}}{y} + 1 \right). \tag{8.5}$$

Theorem 8.1. *The Bond Theorem.*
The price, P, of a noncallable bond immediately following a coupon payment is given by any of the four equivalent equations (8.1), (8.2), (8.4), and (8.5).

If $r = 0$, that is, there are no coupons paid, then the bond is called a ZERO COUPON BOND. Thus, from (8.4) we have the following result.

Theorem 8.2. *The Zero Coupon Bond Theorem.*
The price, P, of a zero coupon bond is given by

$$P = 100(1+y)^{-n}.$$

Comments About the Bond Theorem

- There are three different annual yields used in connection with bonds: the NOMINAL YIELD $r^{(m)} = rm$, the CURRENT YIELD $100rm/P$, and the YIELD TO MATURITY $y^{(m)} = ym$. The yield to maturity is frequently referred to just as the yield. The nominal yield or stated yield, is of little practical significance. The yield to maturity is the annual rate of return that an investor should receive if the bond is held to maturity. The current yield measures that part of the yield to maturity that is due to coupon payments.
- From (8.5), written in the form

$$P - 100 = 100(r-y) \frac{1-(1+y)^{-n}}{y},$$

we see that $P = 100$ is equivalent to $y = r$. If $P = 100$, then the bond is said to be purchased at PAR. In this case $y = r = 100r/P$. Thus, when a bond is purchased at par, the nominal yield, the current yield, and the yield to maturity are equivalent.

- From (8.5), written in the form

$$P - 100 = 100(r - y)\frac{1 - (1+y)^{-n}}{y},$$

and realizing that $(1 + y)^n > 1$, so that $1 - (1+y)^{-n} > 0$, we see that the sign of $P - 100$ is the same as the sign of $r - y$. If $P > 100$ (which is equivalent to $r > y$), then the bond is said to be purchased at a PREMIUM. If $P < 100$ (which is equivalent to $r < y$), then the bond is said to be purchased at a DISCOUNT.

- From (8.5), written in the form

$$y - \frac{100r}{P} = (y - r)\frac{100}{P}(1+y)^{-n},$$

we see that the sign of $y - r$ is the same as the sign of $y - 100r/P$. Thus, if a bond is purchased at a premium (so $P > 100$ and $r > y$), then $y - 100r/P < 0$, so $ym < 100rm/P < rm$, that is, the yield to maturity is less than the current yield, which is less than the nominal yield. However, if a bond is purchased at a discount (so $P < 100$ and $r < y$), then $y - 100r/P > 0$, so $ym > 100rm/P > rm$, that is, the yield to maturity is greater than the current yield, which is greater than the nominal yield.

- The relationship between the yield to maturity, ym, and the maturity,[4] n/m, for bonds with similar degrees of default risk is represented graphically by a yield curve. During economic expansion, yield curves are generally upward sloping, that is, the yields on long-term bonds are higher than the yields on short-term bonds. However, at economic peaks, yield curves often become inverted, that is, the yields on long-term bonds are lower than the yields on short-term bonds.

Example 8.1. What are the semi-annual coupon payments and the price of a 9% noncallable bond with 20 years to maturity and a face value of $1,000 if the yield to maturity is 12%? What are the nominal and current yields for this bond? Is the bond selling at par, at a premium, or at a discount?

Solution. Here the coupons are paid twice a year ($m = 2$), so there are 40 payments ($n = 40$). We also have $F = 1000$ and $r^{(2)} = 0.09$, so $r = 0.045$ and the semi-annual coupon payments, rF, are $0.045 \times 1,000 = \$45$. The yield to maturity is $y^{(2)} = 0.12$, so $y = 0.06$. From (8.5) we find that

$$P = 100\left((0.045 - 0.06)\frac{1 - (1+0.06)^{-40}}{0.06} + 1\right) = 77.43.$$

Thus, the bond should sell for 77.43% of its face value, that is, for $77.43/100 \times 1000 = \$774.30$.

---

[4] We use the terms "maturity" and "years to maturity" interchangeably.

The nominal yield is $rm = 0.045 \times 2 = 0.09$, and the current yield is $100rm/P = 100 \times 0.045 \times 2/77.43 \approx 0.116$. The bond is selling at a discount because $P < 100$. $\triangle$

Example 8.2. What are the semi-annual coupon payments and the price of a 12% noncallable bond with 2 years to maturity and a face value of $1,000 if the yield to maturity is 9%?

Solution. Here the coupons are paid twice a year $(m = 2)$, so there are 4 payments $(n = 4)$. We also have $F = 1000$ and $r^{(2)} = 0.12$, so $r = 0.06$ and the semi-annual coupon payments, rF, are $0.06 \times 1,000 = \$60$. The yield to maturity is $y^{(2)} = 0.09$, so $y = 0.045$. From (8.5) we find that

$$P = 100 \left((0.06 - 0.045) \frac{1 - (1 + 0.045)^{-4}}{0.045} + 1 \right) = 105.381.$$

Thus, the bond should sell for about 105.381% of its face value, that is, for $105.381/100 \times 1000 = \$1,053.81$. $\triangle$

There is another way of looking at this last example, which sheds a different light on bonds. Imagine that we buy this bond for $1,053.81 when the yield is 9% and the coupon rate is 12%. We can think of this as lending someone $1053.81 at 9% per annum. Then after six months we receive a $60 payment, of which $1053.81 \times 0.09/2 = \$47.42$ is interest, and the remainder, $60 - \$47.42 = \$12.58$, is used to reduce the principal. This is deducted from $1053.81, giving $1,041.23, which represents the outstanding principal (the value of the bond) with only three payments left. In this way we can construct the BOND AMORTIZATION SCHEDULE shown in Table 8.1. Notice that, at the fourth and final payment, the value of the bond is $1,000—its face value.

Table **8.1.** Bond Amortization Schedule

Period	Payment	Interest	Principal Repaid	Value
0				$1,053.81
1	$60.00	$47.42	$12.58	$1,041.23
2	$60.00	$46.86	$13.14	$1,028.09
3	$60.00	$46.26	$13.74	$1,014.35
4	$60.00	$45.65	$14.35	$1,000.00

Table 8.2 shows the general situation symbolically in spreadsheet format, where P_i is the value of the bond at period i.

Notice that this schedule does not use (8.4). In fact, it can be used to give an alternative derivation of (8.4), as follows. It is not difficult to see that

$$P_2 = (1 + y) P_1 - rF = (1 + y) ((1 + y) P_0 - rF) - rF$$
$$= (1 + y)^2 P_0 - rF (1 + (1 + y)),$$

Table 8.2. Bond Amortization Schedule—Spreadsheet Format

Period	Payment	Interest	Principal Repaid	Value
0				$\frac{P}{100}F = P_0$
1	rF	yP_0	$rF - yP_0$	$(1+y)P_0 - rF = P_1$
2	rF	yP_1	$rF - yP_1$	$(1+y)P_1 - rF = P_2$
$\vdots$	$\vdots$	$\vdots$	$\vdots$	$\vdots$
$n-1$	rF	yP_{n-2}	$rF - yP_{n-2}$	$(1+y)P_{n-2} - rF = P_{n-1}$
n	rF	yP_{n-1}	$rF - yP_{n-1}$	$(1+y)P_{n-1} - rF = P_n$

and

$$P_3 = (1+y)P_2 - rF = (1+y)^3 P_0 - rF\left(1 + (1+y) + (1+y)^2\right),$$

and finally

$$P_n = (1+y)P_{n-1} - rF = (1+y)^n P_0 - rF\left(1 + (1+y) + \cdots + (1+y)^{n-1}\right),$$

or

$$P_n = (1+y)^n P_0 - rF\frac{(1+y)^n - 1}{y}.$$

However, $P_0 = \frac{P}{100}F$ and $P_n = F$ (why?), so

$$F = (1+y)^n \frac{P}{100}F - rF\frac{(1+y)^n - 1}{y},$$

or

$$(1+y)^n \frac{P}{100} = \left(r\frac{(1+y)^n - 1}{y} + 1\right),$$

from which (8.4) follows.

From (8.4) we can construct Table 8.3 which gives the prices of different noncallable bonds,[5] with coupon rates of 5% and 10%, with 5 and 20 years to maturity, and for yields from 4% to 12%. Notice that the price of 100.00 (that is, the price of the bond that exactly equals the face value) occurs when the yield equals the coupon rate, in agreement with one of the comments we made following the Bond Theorem.

Let's look at this table in detail and make a number of observations.

- First, if we look at the second and fourth columns, then we see that for a fixed yield, the price of a 5-year bond with a coupon rate of 10% is larger than the price of a 5-year bond with a coupon rate of 5%. We see that the same phenomenon is true for the 20-year bonds if we compare the third and fifth columns. This leads us to suspect that, in general, the

---

[5] Unless stated otherwise, all coupons are paid semi-annually, so $m = 2$.

Table 8.3. The Prices of Different Noncallable Bonds

Yield	5%/5 years	5%/20 years	10%/5 years	10%/20 years
4%	104.491	113.678	126.948	182.066
5%	100.000	100.000	121.880	162.757
6%	95.735	88.443	117.060	146.230
7%	91.683	78.645	112.475	132.033
8%	87.834	70.311	108.111	119.793
9%	84.175	63.197	103.956	109.201
10%	80.696	57.102	100.000	100.000
11%	77.387	51.862	96.231	91.977
12%	74.240	47.338	92.640	84.954

price increases with the coupon rate, something that is intuitively obvious (if the coupon rate increases, then so does the value of the coupon so the bond is worth more). Thus, we conjecture that P is an increasing function of r, and we prove this shortly.

- Second, if we look at the second and third columns, then we see that for a fixed coupon rate, the price of a 5-year bond is smaller than the price of a 20-year bond if the coupon rate is larger than the yield, but is larger if the coupon rate is smaller than the yield. We see that the same phenomenon is true for the 10% bonds if we compare the fourth and fifth columns. This leads us to suspect that, in general, the price increases with maturity if the coupon rate is larger than the yield and decreases with maturity if the coupon rate is smaller than the yield. (This is intuitively obvious because if the coupon rate is higher than the yield, the investor is receiving more than required over a longer time, and therefore, should expect to pay more. A similar argument holds when the coupon rate is less than the yield.) Thus, we conjecture that P is an increasing function of n if $r > y$ and is a decreasing function of n if $r < y$. We prove this conjecture shortly.

- Third, looking at the second column, we see that the price decreases as the yield increases. We see the same phenomenon in the third, fourth and fifth columns. This leads us to suspect that, in general, the price and the yield move in opposite directions, something that is also intuitively obvious (if the current interest rates rise, then the prudent investor does not invest in a bond that has a lower interest rate unless the price is right, namely, lower). Thus, we conjecture that P is a decreasing function of y. We prove this conjecture shortly as well.

We now look at these three conjectures, by noting that, from the Bond Theorem on p. 116, the price P is a function of the three independent variables, r, n, and y (or equivalently, $y^{(m)}$ because $y^{(m)} = my$). We look at these three dependencies in turn.

First, if we think of P as a function of r (with n and y fixed), then we see from (8.4) that P is linear in r. Because $(1 + y)^n > 1$, the coefficient of r, namely $(1 - (1+y)^{-n})/y$, is always positive. Thus, P is an increasing function of r, that is, as the coupon rate increases so does the price, exactly as expected. Notice also that as n increases, the slope becomes larger, and $\lim_{n\to\infty} P = 100r/y$.

Plotting P versus r gives a line with positive slope—see Fig. 8.2. From (8.5) we see that if $r = y$, then $P = 100$, so this line passes through the point $(y, 100)$. This tells us that when the coupon rate and the yield coincide, the bond should sell for 100% of the face value, namely the face value. We also note that if $r = 0$, then $P = 100(1+y)^{-n}$, so the line must also pass through the point $\left(0, 100(1+y)^{-n}\right)$. This is the price of a zero coupon bond. Thus, the line joins the points $(y, 100)$ and $\left(0, 100(1+y)^{-n}\right)$.

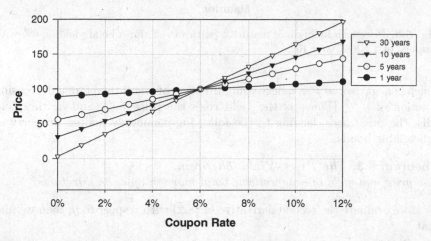

Fig. 8.2. Price as a function of coupon rate for 1, 5, 10, and 30 years with 6% yield

Second, if we think of P as a function of n (with r and y fixed), then we see from (8.5) that P is linear in $-100(r-y)(1+y)^{-n}$. The sequence $\left\{-(1+y)^{-n}\right\}$ is increasing, so P is an increasing function of n if $r > y$ and a decreasing function of n if $r < y$. This is what we expected. We also notice that $\lim_{n\to\infty} P = 100r/y$. This is seen in Fig. 8.3.

Third, we now think of P as a function of y (with r and n fixed). If we differentiate (8.2) with respect to y, then we find that[6]

$$\frac{dP}{dy} = 100r \sum_{k=1}^{n} k(1+y)^{-k-1} - 100n(1+y)^{-n-1},$$

---

[6] Strictly speaking, because P is a function of three independent variables, we should use the notation $\partial P/\partial y$, rather than dP/dy.

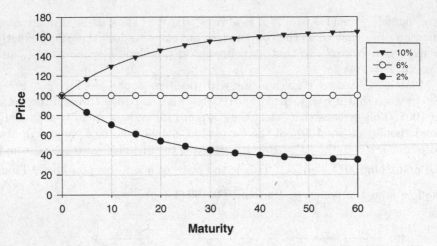

Fig. 8.3. Price as a function of maturity (with $m = 1$) for a bond yielding 6% with coupons of 2%, 6%, and 10%

which is negative, so P is a decreasing function of y and therefore a decreasing function of $y^{(m)}$. Thus, as the yield rises the price falls, and as the yield falls, the price rises, leading to the following commonly quoted property of noncallable bonds.

Theorem 8.3. *The Price-Yield Theorem.*
The price and yield of a noncallable bond move in opposite directions.

If we evaluate the second derivative of (8.2) with respect to y, then we find that

$$\frac{d^2 P}{dy^2} = 100r \sum_{k=1}^{n} k(k+1)(1+y)^{-k-2} + 100n(n+1)(1+y)^{-n-2},$$

which is positive, so the graph of P versus y is concave up. Thus, $P(y)$ is a decreasing, concave up function of y and therefore of $y^{(m)}$ as is demonstrated in Fig. 8.4.

If we think of P as a function of y, and if we are given a value for P, then how do we know that y exists, that is, how do we know that there is always a solution to (8.1)? We know that P is a continuous decreasing function of y for $y \geq 0$. Furthermore

$$\lim_{y \to \infty} P = 0,$$

and from (8.2)

$$\lim_{y \to 0+} P = 100nr + 100.$$

So for there to be a solution, P must satisfy the inequality $0 < P < 100nr + 100$.

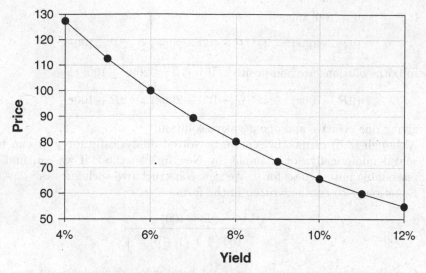

Fig. 8.4. Price versus yield for a 6% bond with a 20-year maturity

Theorem 8.4. *The Yield Existence Theorem.*
If P satisfies

$$0 < P < 100nr + 100,$$

then (8.1) has a solution y for given P, r, and n.

Comments About the Yield Existence Theorem

- By rewriting (8.1) we see that the equation has one change of sign, so the IRR Uniqueness Theorem I on p. 33 applies. Thus, that theorem guarantees that there is no more than one solution for which $1 + y > 0$.
- In general, for $n > 4$, there is no algebraic way to solve (8.1) for y.
- It is instructive to solve (8.1) exactly for $n = 2$ to see where the restriction $0 < P < 200r + 100$ comes into play. In this case we try to solve the quadratic

$$\frac{P}{100}(1+y)^2 = r\left((1+y)+1\right)+1,$$

that is,

$$P(1+y)^2 - 100r(1+y) - 100(1+r) = 0,$$

or

$$Py^2 + y(2P - 100r) + (P - 100 - 200r) = 0$$

for y. This has solutions

$$y = \frac{-(2P - 100r) \pm \sqrt{(2P - 100r)^2 - 4P(P - 100 - 200r)}}{2P}.$$

If $P \geq 200r + 100$, then

$$(2P - 100r)^2 - 4P(P - 100 - 200r) \leq (2P - 100r)^2,$$

so both solutions are non-positive. If $0 < P < 200r + 100$, then

$$(2P - 100r)^2 - 4P(P - 100 - 200r) > (2P - 100r)^2,$$

giving one positive and one negative solution.

- Although (8.5) cannot in general be solved analytically for y, it can be solved numerically, for example, by Newton's method, if we can find a reasonable initial guess for y. We now construct two such guesses. Equation (8.5) can be written in the form

$$y = r - \frac{(P - 100)y}{100\left(1 - (1+y)^{-n}\right)}.$$

Our objective is to simplify the right-hand side's dependence on y using various approximations. If we use the approximation

$$(1+y)^{-n} \approx 1 - ny + \frac{n(n+1)}{2}y^2,$$

valid for y near 0, so that

$$1 - (1+y)^{-n} \approx n\left(y - \frac{(n+1)}{2}y^2\right),$$

then we have

$$y \approx r - \frac{P - 100}{100n\left(1 - \frac{(n+1)}{2}y\right)}.$$

If we use the approximation

$$\frac{1}{1 - \frac{(n+1)}{2}y} \approx 1 + \frac{(n+1)}{2}y,$$

then we have

$$y \approx r - \frac{P - 100}{100n}\left(1 + \frac{(n+1)}{2}y\right).$$

Solving this for y gives

$$y \approx \frac{100nr + 100 - P}{P\frac{n+1}{2} + 100\frac{n-1}{2}}, \tag{8.6}$$

which is used as an approximation for y. If n is large, then this leads to

$$y \approx \frac{\frac{100 - P}{n} + 100r}{\frac{P + 100}{2}}, \tag{8.7}$$

which can be interpreted as follows:

$$y \approx \frac{\text{Average price change per period} + \text{coupon}}{\text{Average of price and face value}}.$$

Either of (8.6) or (8.7) could be used as the initial guess in Newton's Method to find y.

Example 8.3. What is the yield to maturity of a 9% noncallable bond with 20 years to maturity and a face value of $1,000 if the price is 77.43?

Solution. We want to solve (8.5), namely,

$$P = 100 \, (r - y) \, \frac{1 - (1 + y)^{-n}}{y} + 100,$$

for y given $P = 77.43$, $r = 0.09/2 = 0.045$, and $n = 40$, that is,

$$77.43 = 100 \, (0.045 - y) \, \frac{1 - (1 + y)^{-40}}{y} + 100.$$

The two approximations, (8.6) and (8.7), are

$$y \approx \frac{100nr + 100 - P}{P\frac{n+1}{2} + 100\frac{n-1}{2}} = \frac{100 \times 40 \times 0.045 + 100 - 77.43}{77.43 \times \frac{41}{2} + 100 \times \frac{39}{2}} = 0.0573,$$

and

$$y \approx \frac{\frac{100-P}{n} + 100r}{\frac{P+100}{2}} = \frac{\frac{100-77.43}{40} + 100 \times 0.045}{\frac{77.43+100}{2}} = 0.0571,$$

either of which can be used as an initial value in Newton's Method to find that $y = 0.06$, so the yield $y^{(m)}$ is approximately 12%. $\triangle$

Accrued Interest

We have calculated the price of a bond on a coupon date. Now, we look at the case when the bond is sold part way through the coupon period—which is what often happens in practice. We view this through the eyes of the buyer, so the next coupon is the buyer's first coupon, and there are n coupon payment periods remaining. We let w, where $0 < w < 1$, be the proportion of the first coupon payment period remaining when the bond value is calculated. There are two issues to be addressed.

1. The first coupon is owned partly by the seller (for $1 - w$ of a coupon payment period) and partly by the buyer (for w of a coupon payment period).
2. The buyer is going to receive all of the first coupon, so the seller should receive that part of the coupon in proportion to the time that it was technically owned by the seller, namely $(1 - w) rF$, called the ACCRUED COUPON.

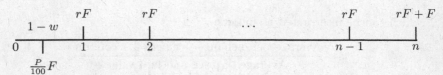

Fig. 8.5. Time diagram

We treat these two issues in turn, using Figure 8.5, which shows the time diagram for this cash flow.

Using the time $1 - w$ as the present time, we find that

$$\frac{P}{100}F = rF\left(\frac{1}{(1+y)^w} + \frac{1}{(1+y)^{1+w}} + \cdots + \frac{1}{(1+y)^{n-1+w}}\right) + \frac{F}{(1+y)^{n-1+w}}.$$

This can be written as follows:

$$\frac{P}{100} = \frac{r}{(1+y)^w}\left[1 + \left(\frac{1}{1+y}\right) + \cdots + \left(\frac{1}{1+y}\right)^{n-1}\right] + \frac{1}{(1+y)^{n-1+w}}$$

$$= \frac{r}{(1+y)^w}\frac{\left(\frac{1}{1+y}\right)^n - 1}{\left(\frac{1}{1+y}\right) - 1} + \frac{1}{(1+y)^{n-1+w}},$$

that is,

$$P = \frac{100r}{(1+y)^w}\left(\frac{1+y-(1+y)^{1-n}}{y}\right) + \frac{100}{(1+y)^{n-1+w}},$$

or

$$P = 100\left((r-y)\frac{1-(1+y)^{-n}}{y} + 1\right)(1+y)^{1-w}.$$

The price that should be paid for the bond is P plus the accrued coupon $(1-w)rF$.

8.2 Duration

Sometimes we need to compare two different bonds, but because the price P is a function of three variables, namely r, y, and n, this is difficult. However, we can compare two bonds from the point of view of risk. In order to introduce this method, we consider the following question. Which bond do you expect to be riskier—a 10-year zero coupon bond or a 10-year noncallable bond, if both have a 7% yield? Most people regard the second bond as carrying less risk than the first, because if the issuers both default, then the second bond has paid some coupons, whereas the first has not.

To quantify this idea, consider placing n weights, $a_1, a_2, \ldots, a_n$, equal distances, s, apart on a weightless beam. We want to find the distance d at which the beam balances. (See Fig. 8.6.)

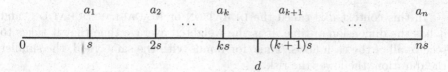

Fig. 8.6. Balancing weights

Assuming that d falls between ks and $(k+1)\,s$, we require that the contributions of the weights to the left of d to balance those on the right, that is,

$$(d - s)\,a_1 + \cdots + (d - ks)\,a_k = ((k+1)\,s - d)\,a_{k+1} + \cdots + (ns - d)\,a_n.$$

Solving this for d gives

$$d = \frac{sa_1 + 2sa_2 + \cdots + nsa_n}{a_1 + a_2 + \cdots + a_n} = s\frac{\sum_{k=1}^{n} ka_k}{\sum_{k=1}^{n} a_k},$$

the weighted average of $a_1, a_2, \ldots, a_n$. In Problem 8.15 you are asked to show that d always lies between s and ns.

We apply this idea to bonds, by considering the future values of the coupons, which are paid m times a year. (See Fig. 8.7.)

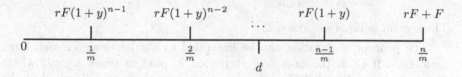

Fig. 8.7. The weighted average of a typical bond

In this case $s = 1/m$, $a_k = rF(1+y)^{n-k}$ for $k = 1, \ldots, n-1$, and $a_n = rF + F$, so[7]

$$d = \frac{1}{m}\frac{\sum_{k=1}^{n-1} krF(1+y)^{n-k} + n(rF + F)}{\sum_{k=1}^{n-1} rF(1+y)^{n-k} + rF + F},$$

---

[7] The same formula results if we consider the present value of the coupons, rather than the future value.

which can be rewritten as

$$d = \frac{1}{m} \frac{r \sum_{k=1}^{n} k(1+y)^{n-k} + n}{r \sum_{k=1}^{n}(1+y)^{n-k} + 1}. \tag{8.8}$$

In this context d is called the DURATION or MACAULAY DURATION, and it has the dimension of time. It is the weighted average time that it takes to receive all of the cash flows. Thus, for bonds with the same yield, the smaller the duration, the lower the risk.

Example 8.4. If the yield to maturity is 6%, then what is the duration of

(a) A 9% noncallable bond with 20 years to maturity?
(b) A 7% noncallable bond with 15 years to maturity?

Which bond is less risky?

Solution.

(a) Here $m = 2$, $r = 0.09/2 = 0.045$, $y = 0.06/2 = 0.03$, and $n = 40$, so from (8.8) we have

$$d = \frac{1}{2} \frac{0.045 \sum_{k=1}^{40} k(1+0.03)^{40-k} + 40}{0.045 \sum_{k=1}^{40}(1+0.03)^{40-k} + 1} = 10.983 \text{ years.}$$

(b) Here $m = 2$, $r = 0.07/2 = 0.035$, $y = 0.06/2 = 0.03$, and $n = 30$, so from (8.8) we have

$$d = \frac{1}{2} \frac{0.035 \sum_{k=1}^{30} k(1+0.03)^{30-k} + 30}{0.035 \sum_{k=1}^{30}(1+0.03)^{30-k} + 1} = 9.787 \text{ years.}$$

The second bond is less risky. $\triangle$

The concept of duration can be extended to any sequence of cash flows as follows. If C_k is the cash flow at period k, paid m times a year, where $0 \le k \le n$, (see Fig. 8.8) then the duration, d, of these cash flows when the interest rate per period is y is the time weighted average of the future values of these cash flows, that is,

$$d = \frac{\sum_{k=0}^{n} \frac{k}{m} C_k(1+y)^{n-k}}{\sum_{k=0}^{n} C_k(1+y)^{n-k}} = \frac{\sum_{k=0}^{n} kC_k(1+y)^{n-k}}{m \sum_{k=0}^{n} C_k(1+y)^{n-k}}.$$

This is equivalent to the time weighted average of the present values of these cash flows,

$$d = \frac{\sum_{k=0}^{n} kC_k(1+y)^{-k}}{m \sum_{k=0}^{n} C_k(1+y)^{-k}}. \tag{8.9}$$

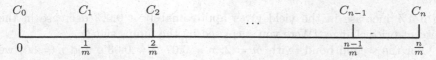

Fig. 8.8. General time diagram

8.3 Modified Duration

We can also compare two bonds by looking at the sensitivity of the prices to a change in yield, that is, if the yield $y^{(m)}$ changes by an amount $\Delta y^{(m)}$, then we can compare how the percent price changes are affected. Thus, we want to calculate the percent price change,[8]

$$\frac{P\left(y^{(m)} + \Delta y^{(m)}\right) - P\left(y^{(m)}\right)}{P\left(y^{(m)}\right)},$$

for each bond. For bonds with the same yield, the larger the magnitude of this number, the greater the risk.

Example 8.5. Which is riskier under a 1% increase in yield, a 9% noncallable bond with 20 years to maturity or a 7% noncallable bond with 15 years to maturity if the yield to maturity in each case is 6%?

Solution. In this case $y^{(m)} = 0.06$ and $\Delta y^{(m)} = 0.01$, so

$$\frac{P(0.06 + 0.01) - P(0.06)}{P(0.06)} = \frac{P(0.07)}{P(0.06)} - 1.$$

For the first bond (with $m = 2$, $r = 0.09/2 = 0.045$, and $n = 40$) we have, from (8.4),

$$P\left(y^{(m)}\right) = 100 \left(0.045 \frac{1 - \left(1 + \frac{y^{(m)}}{2}\right)^{-40}}{\frac{y^{(m)}}{2}} + \left(1 + \frac{y^{(m)}}{2}\right)^{-40} \right),$$

so

$$P(0.06) = 134.672,$$

and

$$P(0.07) = 121.355.$$

Thus, for the first bond

$$\frac{P(0.07)}{P(0.06)} - 1 = -0.0989,$$

---

[8] Although this is called the percent price change, it is not a percent, it is a decimal.

so a 1% increase in the yield gives approximately a 9.9% decrease in the percent price change. (Were you surprised by the minus sign?)

For the second bond (with $m = 2$, $r = 0.07/2 = 0.035$, and $n = 30$) we have, in the same way,

$$P(0.06) = 109.8,$$

and

$$P(0.07) = 100,$$

giving

$$\frac{P(0.07)}{P(0.06)} - 1 = -0.0893,$$

so a 1% increase in the yield gives approximately an 8.9% decrease in the percent price change. (Were you surprised that $P(0.07) = 100$?)

Thus, the first bond is riskier than the second. $\triangle$

We now derive a formula that gives us an approximate measure of the percent price change of a bond. From local linearity, we have

$$P\left(y^{(m)} + \Delta y^{(m)}\right) - P\left(y^{(m)}\right) \approx \Delta y^{(m)} \frac{dP}{dy^{(m)}}$$

for small $\Delta y^{(m)}$, which we can write as

$$\frac{P\left(y^{(m)} + \Delta y^{(m)}\right) - P\left(y^{(m)}\right)}{P\left(y^{(m)}\right)} \approx \Delta y^{(m)} \frac{P'\left(y^{(m)}\right)}{P\left(y^{(m)}\right)}.$$

In this form we see that the percent price change of $P\left(y^{(m)}\right)$ to changes in $y^{(m)}$ is approximately linear in the change, $\Delta y^{(m)}$ and linear in the quantity $P'(y^{(m)})/P\left(y^{(m)}\right)$. Because $P\left(y^{(m)}\right) > 0$ and $P'\left(y^{(m)}\right) < 0$, we see that the sign of the percent price change is the opposite of the sign of the change in $y^{(m)}$.

It is common to define the MODIFIED DURATION by

$$v = -\frac{P'\left(y^{(m)}\right)}{P\left(y^{(m)}\right)}. \tag{8.10}$$

The minus sign is introduced to ensure that v is positive. Modified duration has the dimension of time. (Why?)

With this definition, we have

$$\frac{P\left(y^{(m)} + \Delta y^{(m)}\right) - P\left(y^{(m)}\right)}{P\left(y^{(m)}\right)} \approx -\Delta y^{(m)} v. \tag{8.11}$$

Thus, the percent price change of $P\left(y^{(m)}\right)$ to changes in $y^{(m)}$ is proportional to the modified duration v, once $\Delta y^{(m)}$ is selected.

From (8.11) if $\Delta y^{(m)} = 0.01$, then we have

$$v \approx -100 \left(\frac{P\left(y^{(m)} + 0.01\right) - P\left(y^{(m)}\right)}{P\left(y^{(m)}\right)} \right).$$

We can interpret this as follows: for every 1% increase (or decrease) in the yield, the percent decrease (or increase) in the price is approximately v, the modified duration. For example, the price of a bond with a modified duration of 10.00 decreases by approximately 10% for every 1% increase in the yield.

We derive an equation for v as follows. From (8.4), namely

$$P\left(y^{(m)}\right) = 100 \left(r\frac{1 - (1+y)^{-n}}{y} + (1+y)^{-n} \right)$$

where $y = y^{(m)}/m$, and because

$$\frac{dP}{dy^{(m)}} = \frac{dP}{dy}\frac{dy}{dy^{(m)}} = \frac{dP}{dy}\frac{1}{m},$$

we have

$$\frac{dP}{dy^{(m)}} = 100\frac{1}{m}\left(r\frac{n(1+y)^{-n-1}y - 1 + (1+y)^{-n}}{y^2} - n(1+y)^{-n-1} \right).$$

Thus,

$$v = -\frac{1}{m}\frac{r\frac{n(1+y)^{-n-1}y - 1 + (1+y)^{-n}}{y^2} - n(1+y)^{-n-1}}{r\frac{1 - (1+y)^{-n}}{y} + (1+y)^{-n}}. \tag{8.12}$$

If we multiply the numerator and denominator by $y^2(1+y)^{n+1}$, then we find that

$$v = -\frac{1}{ym}\frac{r\left(ny - (1+y)^{n+1} + (1+y)\right) - ny^2}{r\left((1+y)^{n+1} - (1+y)\right) + y(1+y)},$$

or

$$v = -\frac{1}{ym}\frac{r\left(1 + y + ny - (1+y)^{n+1}\right) - ny^2}{(1+y)\left[r\left\{(1+y)^n - 1\right\} + y\right]}. \tag{8.13}$$

Example 8.6. If the yield to maturity is 6%, then what is the modified duration and the percent price change of

(a) A 9% noncallable bond with 20 years to maturity?
(b) A 7% noncallable bond with 15 years to maturity?

Solution.

(a) Here $m = 2$, $r = 0.09/2 = 0.045$, $y = 0.06/2 = 0.03$, and $n = 40$, so from (8.13), we have

$$v = -\frac{1}{0.03 \times 2} \frac{0.045\left(1 + 0.03 + 40 \times 0.03 - (1 + 0.03)^{40+1}\right) - 40 \times 0.03^2}{(1 + 0.03)\left(0.045\left((1 + 0.03)^{40} - 1\right) + 0.03\right)}$$

$$= 10.663.$$

From (8.11)

$$\frac{P\left(0.06 + \Delta y^{(m)}\right) - P\left(0.06\right)}{P\left(0.06\right)} \approx -10.663 \Delta y^{(m)}.$$

Thus, a 1% increase in the yield gives about a 10.663% decrease in the percent price change compared to the value of 9.89% found in Example 8.5.

(b) Here $m = 2$, $r = 0.07/2 = 0.035$, $y = 0.06/2 = 0.03$, and $n = 30$, so

$$v = -\frac{1}{0.03 \times 2} \frac{0.035\left(1 + 0.03 + 30 \times 0.03 - (1 + 0.03)^{30+1}\right) - 30 \times 0.03^2}{(1 + 0.03)\left(0.035\left((1 + 0.03)^{30} - 1\right) + 0.03\right)}$$

$$= 9.502.$$

From (8.11)

$$\frac{P\left(0.06 + \Delta y^{(m)}\right) - P\left(0.06\right)}{P\left(0.06\right)} \approx -9.502 \Delta y^{(m)}.$$

Thus, a 1% increase in the yield gives about a 9.502% decrease in the percent price change compared to the value of 8.93% found in Example 8.5.

$\triangle$

From (8.2), namely,

$$P = 100r \sum_{k=1}^{n} (1 + y)^{-k} + 100 (1 + y)^{-n},$$

we derive another equation for v that is occasionally useful. If we let

$$p_k = \begin{cases} r(1 + y)^{-k} & \text{for } 1 \le k \le n - 1, \\ (r + 1)(1 + y)^{-n} & \text{for } k = n, \end{cases} \tag{8.14}$$

then

$$P = 100 \sum_{k=1}^{n} p_k.$$

Notice that p_k is the present value of the kth cash flow. From (8.14) we have

$$\frac{dp_k}{dy} = -kp_k (1+y)^{-1} \text{ for } 1 \le k \le n, \tag{8.15}$$

so

$$v = -\frac{\frac{dP}{dy^{(m)}}}{P} = -\frac{1}{m}\frac{\frac{dP}{dy}}{P},$$

giving

$$v = \frac{1}{m}\frac{1}{1+y}\frac{\sum_{k=1}^{n} kp_k}{\sum_{k=1}^{n} p_k}, \tag{8.16}$$

or

$$v = \frac{1}{m}\frac{1}{1+y}\frac{r\sum_{k=1}^{n} k(1+y)^{-k} + n(1+y)^{-n}}{r\sum_{k=1}^{n}(1+y)^{-k} + (1+y)^{-n}}. \tag{8.17}$$

Equations (8.12), (8.13), and (8.17) show that the modified duration is a function of r, y, and n. We investigate each of these dependencies in turn.

We use (8.13) to construct Table 8.4 and Fig. 8.9 for the modified durations of different noncallable bonds with a 9% yield for different coupon rates and maturities.

Table 8.4. The Modified Durations of Different Noncallable Bonds with 9% Yield

Coupon Rate	10 years	20 years	30 years	50 years
3%	7.894	11.744	12.614	11.857
6%	7.030	9.988	10.952	11.200
9%	6.504	9.201	10.319	10.975
12%	6.150	8.755	9.985	10.862

We notice that for a fixed maturity, the modified duration decreases as the coupon rate increases. We prove this as follows. If we differentiate (8.17) with respect to r and simplify, then we find that

$$\frac{dv}{dr} = -\frac{1}{m}\frac{1}{1+y}\frac{\sum_{k=1}^{n}(n-k)(1+y)^{-k-n}}{\left(r\sum_{k=1}^{n}(1+y)^{-k} + (1+y)^{-n}\right)^2}.$$

Because this is negative we have proved that **modified duration is a decreasing function of the coupon rate**. The higher the coupon rate, the faster the investment is repaid, so the lower the modified duration.

We use (8.13) to construct Table 8.5 and Fig. 8.10 for the modified durations of different noncallable bonds with a 9% coupon rate for different yields and maturities.

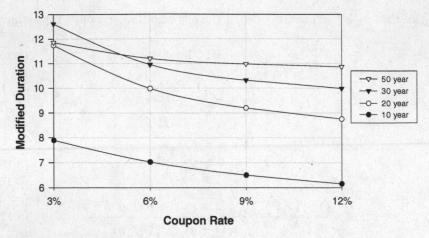

Fig. 8.9. The modified durations of different noncallable bonds with 9% yield

Table 8.5. The Modified Durations of Different Noncallable Bonds with 9% Coupon Rate

Yield	10 years	20 years	30 years	50 years
3%	7.342	12.200	15.988	21.661
6%	6.925	10.663	12.921	15.222
9%	6.504	9.201	10.319	10.975
12%	6.083	7.879	8.283	8.347

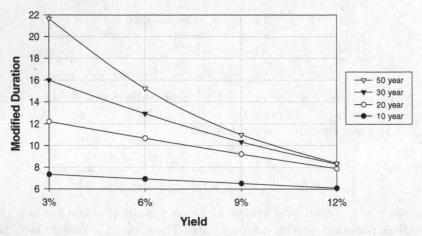

Fig. 8.10. The modified durations of different noncallable bonds with 9% coupon rate

We notice that for a fixed maturity, the modified duration decreases as the yield increases. We prove this as follows. If we differentiate (8.16) with respect to y and simplify, then we find that

$$m\frac{dv}{dy} = -\frac{\left(\sum_{k=1}^n k^2 p_k\right)\left(\sum_{k=1}^n p_k\right) - \left(\sum_{k=1}^n k p_k\right)^2}{(1+y)^2 \left(\sum_{k=1}^n p_k\right)^2} - \frac{\left(\sum_{k=1}^n k p_k\right)}{(1+y)^2 \left(\sum_{k=1}^n p_k\right)}.$$

The second term is negative, so we need concentrate only on the numerator of the first term. If we consider the Cauchy-Schwarz inequality (see p. 250), namely

$$\left(\sum_{k=1}^n a_k b_k\right)^2 \le \left(\sum_{k=1}^n a_k^2\right)\left(\sum_{k=1}^n b_k^2\right),$$

with $a_k = k\sqrt{p_k}$ and $b_k = \sqrt{p_k}$, then we find that

$$\left(\sum_{k=1}^n k p_k\right)^2 \le \left(\sum_{k=1}^n k^2 p_k\right)\left(\sum_{k=1}^n p_k\right).$$

Thus, dv/dy is negative, so **modified duration is a decreasing function of the yield**. The higher the yield, the faster the investment is repaid, so the lower the modified duration.

If we plot the modified duration, v, as a function of the maturity (from 5 to 30 years) for a 9% yield and coupon rates of 3%, 6%, 9% and 12%, then we have Fig. 8.11.

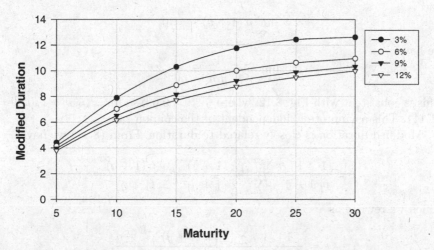

Fig. 8.11. The modified duration as a function of maturity for 9% yield and coupon rates of 3%, 6%, 9% and 12%

This suggests that v is an increasing, concave down function of n. However, if we extend the maturity to 100 years, then we have Fig. 8.12.

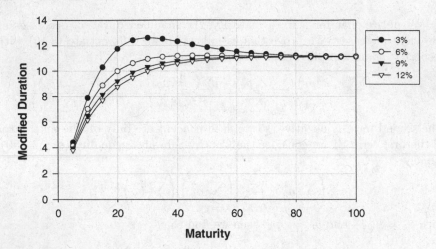

Fig. 8.12. The modified duration as a function of maturity for 9% yield and coupons of 3%, 6%, 9% and 12%

This clearly shows that v is neither an increasing nor a concave down function of maturity.[9] However, it does suggest that there is a horizontal asymptote. So what happens to the modified duration if $n \to \infty$? From (8.12) because

$$\lim_{n\to\infty} (1+y)^{-n} = 0$$

and

$$\lim_{n\to\infty} n (1+y)^{-n} = 0,$$

we have

$$\lim_{n\to\infty} v = \frac{1}{ym} = \frac{1}{y^{(m)}}.$$

This is consistent with Fig. 8.12, where $y^{(m)} = 0.09$, so $\lim_{n\to\infty} v = 1/0.09 = 11.111$. This asymptote is independent of the coupon rate.

Modified duration is closely related to duration. From (8.17) we have

$$v = \frac{1}{m}\frac{1}{1+y}\frac{r\sum_{k=1}^{n} k (1+y)^{-k} + n(1+y)^{-n}}{r\sum_{k=1}^{n} (1+y)^{-k} + (1+y)^{-n}},$$

which we rewrite as

$$v = \frac{1}{1+y}\frac{1}{m}\frac{r\sum_{k=1}^{n} k (1+y)^{n-k} + n}{r\sum_{k=1}^{n} (1+y)^{n-k} + 1}.$$

---

[9] Based on Fig. 8.12 it appears that v is an increasing, concave down function of n if $y \leq r$, while v has a single maximum if $y > r$. You are asked to prove this conjecture for the case $y = r$ (see Problem 8.17). The result for $y \neq r$ has only recently been proved (see [22]).

From (8.8) we find that

$$v = \frac{d}{1+y},$$ (8.18)

so

$$d = -\frac{1}{ym} \frac{r\left(1+y+ny-(1+y)^{n+1}\right)-ny^2}{[r\{(1+y)^n-1\}+y]}.$$ (8.19)

Because $d = (1+y)v$, we use the fact that v decreases with r to state that the **duration is a decreasing function of the coupon rate**. This is intuitively obvious because if the coupon rate increases, then the owner of the bond receives more per coupon period, thereby shortening the duration.

However, from $d = (1+y)v$ and the fact that v decreases with y, we cannot conclude that d decreases with y. (Why?) Nevertheless, in Problem 8.18 you are asked to show that **duration is a decreasing function of yield**. Because duration is the weighted average time that it takes to receive the cash flows, the higher the yield the faster the return of money, hence the shorter the duration.

From $d = (1+y)v$ and

$$\lim_{n\to\infty} v = \frac{1}{y^{(m)}},$$

we have

$$\lim_{n\to\infty} d = \frac{1+y}{y^{(m)}} = \frac{1+\frac{y^{(m)}}{m}}{y^{(m)}}.$$

8.4 Convexity

On p. 130 we saw that for bonds with the same yield, the larger the magnitude of

$$\frac{P\left(y^{(m)}+\Delta y^{(m)}\right)-P\left(y^{(m)}\right)}{P\left(y^{(m)}\right)},$$

the greater the risk. Using local linearity, that is, the first two terms of a Taylor expansion,

$$P\left(y^{(m)}+\Delta y^{(m)}\right)-P\left(y^{(m)}\right) \approx \Delta y^{(m)} P'\left(y^{(m)}\right),$$

we showed in (8.11) that

$$\frac{P\left(y^{(m)}+\Delta y^{(m)}\right)-P\left(y^{(m)}\right)}{P\left(y^{(m)}\right)} \approx -\Delta y^{(m)} v,$$

where the modified duration, v, is

$$v = -\frac{P'\left(y^{(m)}\right)}{P\left(y^{(m)}\right)}.$$

Generally, if $\Delta y^{(m)} > 0$, then because of the minus sign in the definition of v, the larger the modified duration the riskier the bond.

What do we do if two bonds have the same yield and the same modified duration? We include the next term of the Taylor expansion,

$$P\left(y^{(m)} + \Delta y^{(m)}\right) - P\left(y^{(m)}\right) \approx \Delta y^{(m)} P'\left(y^{(m)}\right) + \frac{1}{2}\left(\Delta y^{(m)}\right)^2 P''\left(y^{(m)}\right),$$

so

$$\frac{P\left(y^{(m)} + \Delta y^{(m)}\right) - P\left(y^{(m)}\right)}{P\left(y^{(m)}\right)} \approx -\Delta y^{(m)} v + \frac{1}{2}\left(\Delta y^{(m)}\right)^2 C,$$

where

$$C = \frac{P''\left(y^{(m)}\right)}{P\left(y^{(m)}\right)}.$$

The quantity C is called the CONVEXITY of the bond. So if $\Delta y^{(m)} > 0$ for bonds with the same yield and the same modified duration, then the larger the convexity the riskier the bond.

From (8.3) we can show that the convexity is given by

$$C = \frac{r \sum_{k=1}^{n} k(k+1)\left(1+y\right)^{-k} + n(n+1)\left(1+y\right)^{-n}}{(1+y)^2 \left(r \sum_{k=1}^{n}\left(1+y\right)^{-k} + \left(1+y\right)^{-n}\right)}. \tag{8.20}$$

(See Problem 8.21.)

It is important to realize that the terms "convexity", "second derivative", and "curvature" are not interchangeable. In fact, neither convexity nor the second derivative of a function measure how much a function "bends". The curvature κ of $P\left(y^{(m)}\right)$, namely

$$\kappa = \frac{P''\left(y^{(m)}\right)}{\left(1 + \left(P'\left(y^{(m)}\right)\right)^2\right)^{3/2}},$$

does that. (See Problems 8.22–8.24.)

8.5 Treasury Bills

United States Treasury bills (T-bills) are short term obligations with initial maturities of 4 weeks, 13 weeks, or 26 weeks.[10] They differ from traditional bonds in that they do not pay coupons. They are purchased at less than face value and are redeemed at face value. Thus, a T-bill is discounted from its face value. The profit is the difference between the face value and the purchase price.

Treasury Bills	
Typical Term	4 weeks, 13 weeks, 26 weeks
Payment Frequency	At maturity
Penalty	None
Issuer	Federal Government
Risks	Inflation, Interest Rate, Reinvestment, Maturity
Marketable	Yes
Restrictions	Minimum Investment of $1,000

Table 8.6 shows a sample of Treasury bill quotations from *The Wall Street Journal* of September 1, 2000.

Table 8.6. Treasury Bill Quotations

Maturity	Days to Mat.	Bid	Asked	Chg.	Ask Yld.
Sep 07 2000	6	6.03	5.95	−0.07	6.04
Oct 26 2000	55	6.12	6.08	+0.01	6.22
Jan 04 2001	125	6.07	6.05	+0.01	6.27

The figures in the *Bid* and *Asked* columns are DISCOUNT YIELDS (expressed as percentages) and are calculated using the DISCOUNT YIELD METHOD, which assumes that a year consists of 360 days (that is, twelve 30-day months). If F is the face value of the T-bill, if P is either the purchase price from a dealer (*asked price*) or the selling price to a dealer (*bid price*), and if n is the number of days to maturity, then the discount yield, y_d, is

$$y_d = \frac{F - P}{F} \frac{360}{n} = \frac{(F - P)/F}{n/360}.$$

Thus, the discount yield is the annualized ratio of the profit to the face value of the T-bill, based on a 360-day year. Solving for P gives the price

$$F = F\left(1 - \frac{n}{360}y_d\right).$$

---

[10] This is the situation as of July 1, 2006. Until February 2001, there was also a 52 week T-bill.

Example 8.7. Use Table 8.6 to find the asked price for a \$10,000 T-bill that matures on October 26, 2000.

Solution. Here $F = 10000$, while from Table 8.6 we have $n = 55$ and $y_d = 0.0608$, so

$$P = 10000 \left(1 - \frac{55}{360} 0.0608 \right) = \$9,907.11.$$

$\triangle$

The figures under the *Ask Yld.* column are INVESTMENT YIELDS[11] (expressed as percentages) and are calculated using the INVESTMENT YIELD METHOD. If F is the face value of the T-bill, if P is the purchase price from a dealer (asked price), and if n is the actual number of days to maturity, then the investment yield, y_i, is[12]

$$y_i = \frac{F - P}{P} \frac{365}{n}.$$

Thus, the investment yield is the annualized ratio of the profit to the price of the T-bill, based on a 365-day year. Solving for P gives the price

$$P = \frac{F}{1 + \frac{n}{365} y_i}.$$

The relationship between y_d and y_i is

$$y_i = \frac{365 y_d}{360 - n y_d}.$$

Example 8.8. Use the values of y_d and n from Table 8.6 to confirm the investment yield (*ask yield*) for the T-bill that matures on October 26, 2000.

Solution. From Table 8.6 we have $y_d = 0.0608$ and $n = 55$, so

$$y_i = \frac{365 \times 0.0608}{360 - 55 \times 0.0608} = 0.0622,$$

in agreement with Table 8.6. $\triangle$

From Table 8.6 we see that $y_d < y_i$. Is this true in general?

---

[11] The investment yield is also called the bond equivalent yield, the coupon equivalent yield, the interest yield, and the effective yield.

[12] The reason that the investment yield is calculated using the asked price (as opposed to the bid price) is that the investor buying the bond is interested in the yield, whereas the investor selling the bond is not.

Theorem 8.5. *The T-bill Theorem*
If $P < F$, then $y_d < y_i$.

Proof. Here

$$
\begin{aligned}
y_d &= \frac{F - P}{F} \frac{360}{n} \\
&= \frac{F - P}{P} \frac{P}{F} \frac{365}{n} \frac{360}{365} \\
&= \left(\frac{P}{F} \frac{360}{365} \right) y_i.
\end{aligned}
$$

If $P < F$, then $(P/F)(360/365) < 1$, so $y_d < y_i$. $\qquad\qquad\qquad\square$

8.6 Portfolio of Bonds

It is prudent to diversify investments over several assets. Instead of a single bond, a prudent investor holds a portfolio consisting of several bonds. Thus, we are interested in obtaining a measure of the sensitivity of the value of a portfolio to changes in interest rates. As with a single bond, an appropriate measure of sensitivity for a portfolio is duration.

We assume that the portfolio consists of J bonds, each with maturity n_j/m, price P_j, and duration d_j, where $1 \leq j \leq J$. For each bond the yield per period is y and the number of coupons payments per year is m.

There are two equivalent methods to compute the duration of a portfolio.

In the first method, the duration of the portfolio is the weighted average duration of the bonds that comprise the portfolio, where the weights are the ratios of the market values (prices) of the bonds to the total market value of the portfolio. Thus, if C_{kj} is the cash flow[13] at period k for bond j, then the duration D of the portfolio is

$$
D = \sum_{j=1}^{J} \left(\frac{P_j}{\sum_{h=1}^{J} P_h} \right) d_j = \frac{\sum_{j=1}^{J} P_j d_j}{\sum_{h=1}^{J} P_h},
$$

where

$$
P_j = \sum_{k=0}^{n_j} C_{kj} (1 + y)^{-k}
$$

and

$$
d_j - \frac{\sum_{k=0}^{n_j} k C_{kj} (1 + y)^{-k}}{m \sum_{k=0}^{n_j} C_{kj} (1 + y)^{-k}},
$$

---

[13] For a bond, the cash flow is the coupon payment, which is determined by the face value and the coupon rate.

For example, consider a portfolio consisting of two bonds—Bond 1 (a $10,000 face value, 5% coupon rate, 2-year bond) and Bond 2 (a $20,000 face value, 8% coupon rate, 3-year bond). If the yield is currently 7%, then from (8.4), the market value of Bond 1 is

$$P_1 = 10000 \left(0.025 \frac{1 - (1 + 0.035)^{-4}}{0.035} + (1 + 0.035)^{-4} \right) = \$9,632.69,$$

and the market value of Bond 2 is

$$P_2 = 20000 \left(0.04 \frac{1 - (1 + 0.035)^{-6}}{0.035} + (1 + 0.035)^{-6} \right) = \$20,532.86.$$

Thus, the market value of the portfolio is

$$P_1 + P_2 = \$9,632.69 + \$20,532.86 = \$30,165.55.$$

From (8.19), the duration of Bond 1 is

$$d_1 = -\frac{1}{0.035 \times 2} \frac{0.025 \left(1 + 0.035 + 4 \times 0.035 - (1 + 0.035)^{4+1} \right) - 4 \times 0.035^2}{0.025 \left((1 + 0.035)^4 - 1 \right) + 0.035}$$

$$= 1.926,$$

and the duration of Bond 2 is

$$d_2 = -\frac{1}{0.035 \times 2} \frac{0.04 \left(1 + 0.035 + 6 \times 0.035 - (1 + 0.035)^{6+1} \right) - 6 \times 0.035^2}{0.04 \left((1 + 0.035)^6 - 1 \right) + 0.035}$$

$$= 2.730.$$

Thus, the duration of the portfolio is

$$D = \frac{P_1 d_1 + P_2 d_2}{P_1 + P_2} = \frac{9632.69 \times 1.926 + 20532.86 \times 2.730}{30165.55} = 2.473.$$

In the second method for computing the duration of a portfolio, the cash flows of the bonds that make-up the portfolio are combined and the formula for the computation of duration is applied to this single stream of cash flows according to (8.9). Thus, the duration of the portfolio is

$$\frac{\sum_{k=0}^{n} k (1 + y)^{-k} \sum_{j=1}^{J} C_{kj}}{m \sum_{k=0}^{n} (1 + y)^{-k} \sum_{h=1}^{J} C_{kh}},$$

where n is the maximum of $n_1, n_2, \ldots, n_J$, and $C_{kj} = 0$ if $k > n_j$. In the previous example, the cash flows from Bond 1 are (check these)

Period	1	2	3	4
Cash Flow	$250	$250	$250	$10,250

and those from Bond 2 are (check these)

Period	1	2	3	4	5	6
Cash Flow	$800	$800	$800	$800	$800	$20,800

giving the time diagram shown in Fig. 8.13.

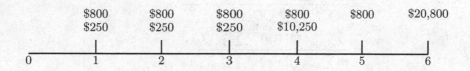

Fig. 8.13. Cash flows of a portfolio

Combining these cash flows gives

Period	1	2	3	4	5	6
Cash Flow	$1,050	$1,050	$1,050	$11,050	$800	$20,800

In order to substitute this into

$$\frac{\sum_{k=0}^{n} k (1+y)^{-k} \sum_{j=1}^{J} C_{kj}}{m \sum_{k=0}^{n} (1+y)^{-k} \sum_{h=1}^{J} C_{kh}},$$

we construct the following table.

k	$\sum_{j=1}^{J} C_{kj}$	$(1+y)^{-k} \sum_{j=1}^{J} C_{kj}$	$k (1+y)^{-k} \sum_{j=1}^{J} C_{kj}$
1	$1,050.00	$1,014.49	$1,014.49
2	$1,050.00	$980.19	$1,960.37
3	$1,050.00	$947.04	$2,841.12
4	$11,050.00	$9,629.44	$38,517.75
5	$800.00	$673.58	$3,367.89
6	$20,800.00	$16,920.81	$101,524.88
Total		$30,165.55	$149,226.50

From this we find that

$$\frac{\sum_{k=0}^{n} k (1+y)^{-k} \sum_{j=1}^{J} C_{kj}}{m \sum_{k=0}^{n} (1+y)^{-k} \sum_{h=1}^{J} C_{kh}} = \frac{149226.50}{2 \times 30165.55} = 2.473,$$

in agreement with the first method.

That the two methods are equivalent is seen from the following.

$$
\begin{aligned}
D &= \frac{\sum_{j=1}^{J} P_j d_j}{\sum_{h=1}^{J} P_h} \\
&= \frac{\sum_{j=1}^{J} \left(\sum_{k=0}^{n} C_{kj} (1+y)^{-k} \right) \frac{\sum_{k=0}^{n} k C_{kj} (1+y)^{-k}}{\sum_{k=0}^{n} C_{kj} (1+y)^{-k}}}{m \sum_{h=1}^{J} \sum_{k=0}^{n} C_{kh} (1+y)^{-k}} \\
&= \frac{\sum_{j=1}^{J} \sum_{k=0}^{n} k C_{kj} (1+y)^{-k}}{m \sum_{h=1}^{J} \sum_{k=0}^{n} C_{kh} (1+y)^{-k}} \\
&= \frac{\sum_{k=0}^{n} k (1+y)^{-k} \sum_{j=1}^{J} C_{kj}}{m \sum_{k=0}^{n} (1+y)^{-k} \sum_{h=1}^{J} C_{kh}}.
\end{aligned}
$$

8.7 Problems

Walking

8.1. Construct a bond amortization schedule, similar to that shown in Table 8.1 on p. 118, for a 9% noncallable bond with 2 years to maturity and a face value of $1,000 if the yield to maturity is 12% and the coupons are paid semi-annually.

8.2. How can you tell from Table 8.4 on p. 133 that modified duration is not an increasing function of maturity?

8.3. How can you tell from Fig. 8.9 on p. 134 that modified duration is not an increasing function of maturity?

8.4. Show that $\lim_{n \to \infty} y^{(m)} = 100 rm/P$, the current yield. Under what circumstances is the current yield of a bond a good approximation for the yield to maturity?

8.5. Confirm that the yields y_i in the *Ask Yld.* column in Table 8.6 on p. 139 are accurate.

8.6. Prove the T-bill Theorem on p. 141 using

$$
y_i = \frac{365 y_d}{360 - n y_d}.
$$

8.7. Use Table 8.6 on p. 139 to find the annual effective rate (EFF) for the T-bill that matures on October 26, 2000.

8.8. By looking at two successive editions of *The Wall Street Journal*, decide what the figures in the *Chg.* column in Table 8.6 on p. 139 represent.

8.9. On March 1, 2000 Hugh Kendrick invests $2,000 in Bond 1 and $2,000 in Bond 2. On the same day Tom Kendrick invests $2,000 in Bond 3 and $8,000 in Bond 4. At the end of six months Bond 1 is worth $2,050, Bond 2 is worth $2,100, Bond 3 is worth $2,040, and Bond 4 is worth $8,360. Confirm that Hugh has a higher (internal) rate of return on each of his bonds than Tom, but Tom has a higher (internal) rate of return on his total portfolio.[14]

8.10. Tom invests $2,000 in a bond, and at the end of one year its value is $2,100. How much must he invest in a second bond that yields 10% over the same period to have a total internal rate of return of 8% on this portfolio?

8.11. If Tom invests $2,000 in a bond with an annual rate of return of 4.5% and $4,000 in a second bond with the same maturity, then what rate of return is needed on the second bond to have a total internal rate of return of 8% on this portfolio?

8.12. Wendy and Amanda Kendrick purchase bonds on the same day. Both bonds have a yield of 6% and a face value of $10,000.

(a) If Wendy's bond is a 9% bond with 15 years to maturity, then how much does she pay for it?
(b) If Amanda's bond has 10 years to maturity and she pays $11,487.75 for it, then what is the coupon rate?

8.13. Referring to Problem 8.12, find the duration and modified duration for both bonds.

8.14. Referring to Problems 8.12 and 8.13, find the duration and modified duration for the single portfolio consisting of both Wendy's bond and Amanda's bond.

Running

8.15. Prove that if $a_1, a_2, \cdots, a_n$ are positive, then

$$\sum_{k=1}^{n} a_k \le \sum_{k=1}^{n} k a_k \le n \sum_{k=1}^{n} a_k.$$

Use this to show that if

$$d = s \frac{\sum_{k=1}^{n} k a_k}{\sum_{k=1}^{n} a_k},$$

then $s \le d \le ns$. (See p. 127.)

8.16. Show that the duration can never exceed the maturity, that is, $d \le n/m$.

---

[14] This is an example of Simpson's paradox, which states that if $a/b > c/d$ and $e/f > g/h$, then it need not follow that $(a+e)/(b+f) > (c+g)/(d+h)$.

8.17. Show that if a bond is selling at par, that is, $r = y$, then the modified duration, v, is an increasing, concave down function of maturity, n/m. (See p. 136.)

8.18. If p_k are given by (8.14), then show that

$$md = \frac{\sum_{k=1}^n kp_k}{\sum_{k=1}^n p_k}.$$

Differentiate this with respect to y and use the Cauchy-Schwarz inequality to show that duration is a decreasing function of yield. (See p. 137.)

8.19. Show that for a zero coupon bond, we have $d = (1/m)n$, which is a line of slope 1 if duration is plotted as a function of maturity. Conversely, show that if $d = (1/m)n$, then $r = 0$ or $n = 1$, a zero coupon bond.

8.20. Suppose that the price of an investment is $\sum_{k=1}^n C/(1+y)^k$, the cash flow $C > 0$ is paid over n time periods, and the yield is $y > 0$. Thus, the total of the cash flows is nC. Compare this to a second investment in a zero coupon bond with yield y and face value nC, which is paid at time T.

(a) Show that the price of the second investment is $nC/(1+y)^T$.
(b) Show that, in order to have the prices of the two investments equal, we need

$$(1+y)^{-T} = \sum_{k=1}^n \frac{(1+y)^{-k}}{n}.$$

(c) Prove that if T satisfies the equation in part (b), then $T \le (n+1)/2$.
(d) Express T in terms of y and n.
(e) Show that, with n fixed, T is a decreasing function of y.
(f) Show that, with n fixed, $\lim_{y \to 0^+} T = (n+1)/2$.

8.21. Show, from (8.3) on p. 115, that the convexity, C, of a bond is given by

$$C = \frac{r \sum_{k=1}^n k(k+1)(1+y)^{-k} + n(n+1)(1+y)^{-n}}{(1+y)^2 \left(r \sum_{k=1}^n (1+y)^{-k} + (1+y)^{-n} \right)}.$$

(See (8.20) on p. 138.)

8.22. If $P(y) = y^2$, then evaluate

(a) $P''(y)$.
(b) $P''(y)/P(y)$.
(c) $P''(y)/ \left(1 + (P'(y))^2 \right)^{3/2}$.

Does the graph of the parabola $P(y) = y^2$ have a constant "bend" for all y? Which of your answers to parts (a), (b), and (c) is constant? Is that a reasonable measure of the "bend" of the function? (See p. 138.)

8.23. If $P(y) = e^y$, then evaluate

(a) $P''(y)$.
(b) $P''(y)/P(y)$.
(c) $P''(y)/\left(1 + (P'(y))^2\right)^{3/2}$.

Does the graph of the exponential function $P(y) = e^y$ have a constant "bend" for all y? Which of your answers to parts (a), (b), and (c) is constant? Is that a reasonable measure of the "bend" of the function? (See p. 138.)

8.24. If $P(y) = \sqrt{1 - y^2}$, then evaluate

(a) $P''(y)$.
(b) $P''(y)/P(y)$.
(c) $P''(y)/\left(1 + (P'(y))^2\right)^{3/2}$.

Does the graph of the semi-circle $P(y) = \sqrt{1 - y^2}$ have a constant "bend" for all y? Which of your answers to parts (a), (b), and (c) is constant? Is that a reasonable measure of the "bend" of the function? (See p. 138.)

Questions for Review

- What is the difference between a callable and a noncallable bond?
- What is the yield to maturity? The current yield?
- How do you calculate the price of a bond immediately following a coupon payment?
- What is a zero coupon bond?
- What does it mean to say that a bond is purchased at a premium? At a discount?
- How is the price of a bond related to r? To n? To y?
- What is duration?
- How do you compute the duration of a bond?
- How is duration related to r? To y?
- How do you compute the duration of a portfolio?
- What is modified duration?
- How do you compute the modified duration of a bond?
- How is modified duration related to r? To y?
- What do the Bond and Zero Coupon Bond Theorems say?
- What is the Price-Yield Theorem?
- What is the Yield Existence Theorem?
- What is a T-bill?
- How are the discount and investment yields for a T-bill computed?
- What does the T-bill Theorem say?

9

Stocks and Stock Markets

Stock represents ownership of a corporation. The stockholders, or shareholders, are the holders of the stock. There are different types of stock, but in this book we study common stock.[1]

A corporation in need of funds may issue stock to private investors. The investors purchase shares of stock in the company. These investors assume a large amount of risk in return for the possibility of growth of the company and a corresponding increase in the value of the shares. If the company is successful, then it may decide to "go public" and offer shares of stock to the general public. This is accomplished through an initial public offering. If in the future the company wants to raise additional funds, then it may have a secondary public offering. It is important to realize that the corporation receives its money when the shares are issued. Any trading after that point takes place between the shareholders and the persons wishing to purchase the stock, and does not directly represent a profit or loss to the company.

Investors who purchase stock may receive dividends periodically (usually quarterly). Thus, the investor may profit in two ways: through an increase in the value of the stock and through the receipt of dividends. There can be substantial risk for the shareholders, but historically stocks have been a very good investment for individuals who hold stocks for long periods of time.

Corporations sometimes use stock splits to create more shares of the stock at a lower price per share. For example, if a company declares a two-for-one stock split, then each shareholder receives two new shares for each old share of the stock. The price of each new share is initially one-half the price of an old share. Companies may do this to attract new buyers since the new price is less than the old price. Less frequently, a company may declare a reverse

---

[1] Owners of common stock have voting rights and are entitled to the earnings of the company after all obligations are paid.

stock split. For example, it may declare a two-for-one reverse split. In this case one new share is issued for every two old shares. A company may do this to attract institutional investors, who may have a minimum price requirement per share.

The first stock market in the United States originated in Philadelphia in 1790. It eventually became known as the Philadelphia Stock Exchange. The New York Stock Exchange (NYSE), initiated in 1792, is the largest stock exchange in the world. As many as one billion shares are traded during a single day on the NYSE. Trading on this exchange occurs auction style. The buyers and sellers have the option of sending bids and offers to the exchange and accepting the bids and offers from others at the exchange. Stock is sold to the highest bidder and bought at the lowest offer. The NASDAQ (National Association of Securities Dealers Automatic Quotation System) is another major United States stock exchange. As with the NYSE, over one billion shares have been traded on this exchange during some trading days. This exchange differs in operation from the NYSE in that orders are placed and trades are made electronically.

There are several types of risk involved in the purchase and sale of stock. Among these are economic risk, interest rate sensitivity, the possibility of company failure, company management problems, competition from other companies, and governmental rulings that may negatively affect the company. In order to study some of these risks, we break companies into three different groups, depending on the CAPITALIZATION—the total value of issued shares[2]—of the company.[3]

Common Stock
Issuer Corporation
Risks Default, Exchange Rate, Interest Rate, Market Price, Volatility, Political

The largest companies are called big caps or large caps. In general, these companies have a very high capitalization (over $10 billion). These are large companies with an established track record. There is usually little possibility of company failure. Stock prices are usually relatively high, and thus there may not be a high growth potential. Many of these companies pay regular dividends.

Mid-cap companies have a capitalization of around $1.5 billion to $10 billion. These companies usually have a higher growth potential than big cap companies.

Small-cap companies have a capitalization of less than $1.5 billion. These companies have the highest growth potential but also have a corresponding higher chance of company failure.

---

[2] The total value of issued shares is the product of the share price and the number of shares issued. Shares that have been issued to investors are frequently called "outstanding shares".

[3] The capitalization cut-off values between these groups are approximate.

9.1 Buying and Selling Stock

In order to buy or sell stock in the United States, the investor uses an investment firm registered with the appropriate governmental agencies—the Securities and Exchange Commission (SEC) and the National Association of Securities Dealers (NASD). The person actually making the transactions for the firm is called a broker.[4]

The fee or commission charged for the transaction is an important consideration. Fees are highest for a full-service broker, who can also provide investment advice. Investors placing trades through the Internet usually pay lower commissions. It is important to note that a commission is charged for both buying and selling stock.

There are different ways to buy and sell stock. The most common way is to pay the full price in cash. For example, if Helen Kendrick buys 500 shares of a stock when the price is $20 a share, then she pays $10,000 to her investment firm plus the commission for the purchase. If the stock increases in price to $30 per share, then she sells the stock through her investment firm, receiving $15,000 less commission for the sale. So Helen makes a profit of $5,000 less any commissions.

Suppose that Helen and Hugh Kendrick both buy the same stock on a regular monthly basis using different methods. Helen buys 10 shares every month, while Hugh buys $100 worth of the same shares, regardless of the number that can be purchased for that amount. The method of investing used by Hugh is called DOLLAR COST AVERAGING. We discuss these two methods in more detail.

Let $S(t)$ be the stock price at time t. If we buy the same number of shares, N, at times $t = 1$ and $t = 2$, then we pay $NS(1) + NS(2)$ for a total of $2N$ shares, so the average price per share is

$$\frac{NS(1) + NS(2)}{2N} = \frac{S(1) + S(2)}{2} = \frac{1}{2}\sum_{t=1}^{2} S(t).$$

If we buy the same number of shares at times $t = 1, \ldots, n$, then the average price per share is

$$\frac{1}{n}\sum_{t=1}^{n} S(t).$$

If we use dollar cost averaging and spend D dollars to buy shares at times $t = 1$ and $t = 2$, then we pay $2D$ for a total of $D/S(1) + D/S(2)$ shares, so the average price per share is

$$\frac{2D}{\frac{D}{S(1)} + \frac{D}{S(2)}} = \frac{2}{\frac{1}{S(1)} + \frac{1}{S(2)}} = \frac{2}{\sum_{t=1}^{2}\frac{1}{S(t)}}$$

---

[4] We use the terms "the firm" and "the broker" interchangeably.

If we spend D dollars to buy shares at times $t = 1, \ldots, n$, then the average price per share is

$$\frac{n}{\sum_{t=1}^{n} \frac{1}{S(t)}}.$$

The question we want to answer is, "Which method gives the lower average price per share?"

First, we concentrate on $n = 2$. If we look at the difference between the average prices per share, then we see that

$$
\begin{aligned}
\frac{S(1) + S(2)}{2} - \frac{2}{\frac{1}{S(1)} + \frac{1}{S(2)}} &= \frac{S(1) + S(2)}{2} - \frac{2S(1)S(2)}{S(1) + S(2)} \\
&= \frac{(S(1) + S(2))^2 - 4S(1)S(S)}{2(S(1) + S(2))} \\
&= \frac{(S(1) - S(2))^2}{2(S(1) + S(2))}.
\end{aligned}
$$

This difference is positive unless $S(1) = S(2)$, in which case it is zero. Thus, unless $S(1) = S(2)$, dollar cost averaging gives the lower average share price if $n = 2$. If $S(1) = S(2)$, then both methods give the same average price per share.

This can be generalized to n time intervals by considering the sign of the quantity

$$\frac{1}{n} \sum_{t=1}^{n} S(t) - \frac{n}{\sum_{t=1}^{n} \frac{1}{S(t)}} = \frac{1}{n \sum_{t=1}^{n} \frac{1}{(t)}} \left(\left(\sum_{t=1}^{n} S(t) \right) \left(\sum_{t=1}^{n} \frac{1}{S(t)} \right) - n^2 \right).$$

Now, the Cauchy-Schwarz inequality (see Appendix A.3 on p. 249) states that

$$\left(\sum_{t=1}^{n} a_t b_t \right)^2 \leq \left(\sum_{t=1}^{n} a_t^2 \right) \left(\sum_{t=1}^{n} b_t^2 \right),$$

with equality if and only if either $a_t = \lambda b_t$ $(t = 1, 2, \ldots, n)$ for some constant λ, or $b_t = 0$ $(t = 1, 2, \ldots, n)$. With $a_t = \sqrt{S(t)}$ and $b_t = 1/\sqrt{S(t)}$, we see that

$$n^2 \leq \left(\sum_{t=1}^{n} S(t) \right) \left(\sum_{t=1}^{n} \frac{1}{S(t)} \right),$$

with equality if and only if $\sqrt{S(t)} = \lambda/\sqrt{S(t)}$, that is, $S(1)) = S(2) = \cdots = S(n) = \lambda$. Thus, unless $S(1) = S(2) = \cdots = S(n)$, then dollar cost averaging gives the lower average price per share. If $S(1) = S(2) = \cdots = S(n)$, then both methods give the same average price per share.

This leads to the following theorem.

Theorem 9.1. *The Dollar Cost Averaging Theorem.*
If we are buying shares of stock, then dollar cost averaging gives a lower average price per share than buying a fixed number of shares on a regular basis. If we are selling shares, then selling a fixed number of shares on a regular basis gives a higher average price per share than dollar cost averaging.

Another way to buy stock is BUYING ON MARGIN, which allows investors to leverage their purchases. An investor who wishes to buy stock on margin must set up a margin account with a broker. The investor may then purchase stock in this account using money borrowed from the broker. However, the investor is required to deposit a down payment on the purchase of the stock. The required minimum down payment, the **margin equity**, is 50% of the purchase price of the stock.[5] This is called the **margin requirement**. The **equity** in the margin account is the difference between the current market value of the shares and the money owed to the broker for the purchase of the stock, the **debit balance**. As the market value of the shares fluctuates then so will the equity in the margin account. Although the margin requirement applies only to the initial purchase of the stock, the investor is required to maintain sufficient equity in the margin account at all times. The required minimum equity, the **required (maintenance) equity**, is a specified proportion, the **maintenance level**, of the current market value of the shares. This is called the **maintenance requirement**.

We illustrate this with the following example. An investor opens a margin account with a deposit of $5,000 and borrows $5,000 from the broker, so the debit balance is $5,000. The investor uses the $10,000 to buy 500 shares of stock at $20 a share. At this point the investor has a total of $10,000−$5,000 = $5,000 in equity. In this example, the maintenance requirement is 25% of the market value of the shares in the account. This means that the equity in the account must be at least 25% of the market value of the shares. The required equity is $0.25 \times \$10,000 = \$2,500$, which is smaller than the equity of $5,000, so the investor satisfies the maintenance requirement.

This is expressed symbolically in Table 9.1, where m is the maintenance level (expressed as a decimal).

Table 9.1. Buying on Margin

Price	Number Shares	Market Value	Debit Balance	Equity	Required Equity	Margin Equity
S	N	$V = NS$	D	$E = V - D$	mV	$0.5V$
$20	500.00	$10,000.00	$5,000.00	$5,000.00	$2,500.00	$5,000.00

[5] The rate of 50% is the rate in effect in the United States as of July, 2006.

We consider three scenarios.

1. The price of the stock rises to $30 per share.
   (a) If the investor sells all of the shares, then the investor makes a profit of $500 \times (\$30 - \$20) = \$5,000$ less any commissions and interest charged for the loan.
   (b) If the investor sells no shares, then because the market value of the shares is now $V = NS = 500 \times \$30 = \$15,000$, but the debit balance is still $D = \$5,000$, the equity is now $E = V - D = \$15,000 - 5,000 = \$10,000$. The required equity is $mV = 0.25 \times \$15,000 = \$3,750$, so the maintenance requirement is satisfied. The margin equity is $0.5V = 0.5 \times \$15,000 = \$7,500$. Thus, there is excess equity in the margin account of $\$10,000 - \$7,500 = \$2,500$.

   This excess equity can be used to borrow an additional $\$2,500/0.5 = \$5,000$ to purchase $\$5,000/\$30 = 166.67$ additional shares. The market value of the shares is now $V = 666.67 \times \$30 = \$20,000$, and the investor has a debit balance of $D = \$5,000 + \$5,000 = \$10,000$.[6] The equity is $E = V - D = \$20,000 - \$10,000 = \$10,000$, the required equity is $0.25V = 0.25 \times \$20,000 = \$5,000$, and the margin equity is $0.5V = 0.5 \times \$20,000 = \$10,000$. As long as the stock price rises, the investor can buy additional shares by borrowing from the broker without investing more money. This is an example of leveraging money. Table 9.2 follows this example as the stock price increases from $20 to $50 a share assuming that the excess equity is always used to purchase additional shares on margin.

Table 9.2. The Effect of a Price Increase When Stock is Purchased on Margin

Price	Number Shares	Market Value	Debit Balance	Equity	Required Equity	Margin Equity
$20	500.00	$10,000.00	$5,000.00	$5,000.00	$2,500.00	$5,000.00
$30	500.00	$15,000.00	$5,000.00	$10,000.00	$3,750.00	$7,500.00
$30	666.67	$20,000.00	$10,000.00	$10,000.00	$5,000.00	$10,000.00
$40	666.67	$26,666.67	$10,000.00	$16,666.67	$6,666.67	$13,333.33
$40	833.33	$33,333.33	$16,666.67	$16,666.67	$8,333.33	$16,666.67
$50	833.33	$41,666.67	$16,666.67	$25,000.00	$10,416.67	$20,833.33
$50	1000.00	$50,000.00	$25,000.00	$25,000.00	$12,500.00	$25,000.00

At this stage the investor owns 1,000 shares with a market value of $50,000, and the equity is $25,000. The value of the investor's position in the stock has quintupled! It is important to note that the investor pays commissions on each purchase and pays interest on the loans.

---

[6] Note that at this point the broker loaned the investor another $5,000 for the purchase of additional shares, bringing the debit balance to $10,000.

2. The price of the stock falls to $15 per share. At this point the required equity is $mV = mNS = 0.25 \times 500 \times \$15 = \$1,875$, and the equity is $E = V - D = 500 \times \$15 - \$5,000 = \$2,500$, so the maintenance requirement is satisfied. These calculations are summarized in Table 9.3.

Table 9.3. The Effect of a Price Decrease When Stock is Purchased on Margin

Price	Number Shares	Market Value	Debit Balance	Equity	Required Equity	Margin Equity
$20	500.00	$10,000.00	$5,000.00	$5,000.00	$2,500.00	$5,000.00
$15	500.00	$7,500.00	$5,000.00	$2,500.00	$1,875.00	$3,750.00

3. The price of the stock falls to $10 per share. At this point the required equity is $mV = 0.25 \times 500 \times \$10 = \$1,250$, and the equity is $E = V - D = 500 \times \$10 - \$5,000 = \$0$. The value of the shares equals the amount owed for the purchase of the stock! Thus, the maintenance requirement is not met, and the investor receives a **maintenance call** from the broker because the investor owes $1,250. To meet the maintenance requirement the investor is required to deposit $1,250 into the margin account, increasing the equity in the account so that the maintenance requirement is met.[7] If the price continues to fall, then the investor is required to deposit additional money to meet the maintenance requirement. It should be noted that the investor's potential loss is limited to the original price of the shares plus commissions plus interest on the loan. Table 9.4 follows this example as the stock price decreases from $20 to $5 a share.

Table 9.4. The Effect of a Price Decrease When Stock is Purchased on Margin

Price	Number Shares	Market Value	Debit Balance	Equity	Required Equity	Margin Equity
$20	500.00	$10,000.00	$5,000.00	$5,000.00	$2,500.00	$5,000.00
$15	500.00	$7,500.00	$5,000.00	$2,500.00	$1,875.00	$3,750.00
$10	500.00	$5,000.00	$5,000.00	$0.00	$1,250.00	$2,500.00
$10	500.00	$5,000.00	$3,750.00	$1,250.00	$1,250.00	$2,500.00
$5	500.00	$2,500.00	$3,750.00	-$1,250.00	$625.00	$1,250.00
$5	500.00	$2,500.00	$1,875.00	$625.00	$625.00	$1,250.00

At this point the investor has paid a total of $3,125 in maintenance fees[8] to meet the maintenance requirement. This is in addition to the initial

---

[7] The investor could also sell some shares.

[8] When the price falls to $10 a share, the investor pays $1,250. When it falls to $5 a share the equity is $-\$1,250$, and the required equity is $625, so the investor pays $\$625 + \$1,250 = \$1,875$, for a total of $\$1,250 + \$1,875 = \$3,125$.

deposit of \$5,000. If the stock price continues to fall, then the investor is required to deposit more money, but the total amount of the maintenance fees can never exceed \$5,000, the original debit balance.

Notice from Table 9.4 that somewhere between a share price of \$15 and a share price of \$10, there must be a price, S, at which the required equity equals the equity, below which a maintenance call is issued. In this case that will occur when

$$0.25 \times 500S = 500S - 5000,$$

which gives $S = \$13.33$.

A natural question to ask is, "To what level can the stock price fall before a maintenance call is issued?" This is answered in the following theorem.

Theorem 9.2. *The Long Sale Maintenance Level Theorem.*
If a stock is purchased on margin, if the original market value of the shares purchased is V_o, if the original equity is E_o, and if the maintenance level is m, then the minimum market value, V^*, for which the equity, E, is equal to or greater than the required equity (and so does not generate a maintenance call) is

$$V^* = \frac{1}{1-m}(V_o - E_o).$$

Proof. We notice that $D = V_o - E_o$ is the debit balance, the money owed to the broker for the purchase of the stock. However, if V is the current market value of the shares and E is the current equity, then $V - E$ is also the money owed to the broker for the purchase of the stock (why?), so

$$V - E = V_o - E_o.$$

At the minimum market value V^* we have $E^* = mV^*$, so

$$V^* - mV^* = V_o - E_o,$$

that is,

$$V^* = \frac{V_o - E_o}{1-m}.$$

We now show that $(V_o - E_o)/(1-m)$ is the minimum market value.

If $V < (V_o - E_o)/(1-m)$, then the current market value is less than $1/(1-m)$ times the money owed to the broker for the purchase of the stock. This means that $V < (V - E)/(1-m)$, which gives $E < mV$, a maintenance call. $\qquad\square$

Example 9.1. To avoid a maintenance call, what is the minimum level to which the market value of the stock can fall in a margin account containing 500 shares if the original market value is \$10,000, if the original equity is \$5,000, and if the maintenance level is 0.25? At what price per share does this occur?

Solution. Here $V_o = 10000$, $E_o = 5000$, and $m = 0.25$, so the minimum market value of the stock is $1/0.75 \times (\$10,000 - \$5,000) = \$6,666.67$. The required equity is $0.25 \times \$6,666.67 = \$1,666.67$, and the equity is $\$6,666.67 - \$5,000 = \$1,666.67$. At this level the price per share of the stock is $\$6,666.67/500 = \$13.33$, in agreement with our previous result. $\triangle$

If an investor believes that a stock is going to decrease in value, then the investor may borrow shares from the broker, and then sell the stock. This is called SHORT SELLING. If the stock decreases in value, then the investor may purchase an equivalent number of shares at the lower price, and use those shares to repay the loan.

The following example illustrates this. An investor wishes to short sell 100 shares of stock, which is currently selling at $50 per share. The market value of the stock is $100 \times \$50 = \$5,000$. The margin requirement is 50% of the market value of the stock, so the investor deposits $0.50 \times \$5,000 = \$2,500$ in a margin account, and the broker loans the investor 100 shares. The investor then sells the stock for $50 per share, and receives $100 \times \$50 = \$5,000$ (less commissions). So the investor has a credit balance of $\$2,500 + \$5,000 = \$7,500$ (the sum of the proceeds from the sale plus the initial deposit).

Because the investor must return the loaned shares to the broker, if there is an increase in price, then the investor loses money on the stock, and the investment firm incurs a risk. Therefore, there is a maintenance requirement for short sales. We assume that the maintenance requirement is 30%, which in this example is $0.3 \times \$5,000 = \$1,500$. This means that the equity in the account must be at least 30% of the value of the shares at all times. The equity is the difference between the credit balance and the current market value of the stock, so in this example the equity is $\$7,500 - \$5,000 = \$2,500$, and the maintenance requirement is satisfied.

This is shown symbolically in Table 9.5, where m is the maintenance level (expressed as a decimal).

Table 9.5. Selling Stock Short

Credit Balance	Price Per Share	Number of Shares	Market Value	Equity	Required Equity
C	S	N	$V = NS$	$E = C - V$	mV
$7,500	$50	100	$5,000	$2,500	$1,500

We consider two cases.

1. The price of the stock falls to $40 per share.
   (a) If the investor buys the stock at $40 per share, and returns the shares to the firm, then the investor makes a profit of $100 \times (\$50 - \$40) = \$1,000$, less commissions on the sale and purchase of the shares and any interest due to the broker on the loan of the shares.

(b) If the investor does not buy the stock at this time, then because the market value of the stock is $40 \times 100 = \$4,000$, the required equity is 30% of \$4,000, that is, $0.3 \times \$4,000 = \$1,200$. The equity is $\$7,500 - \$4,000 = \$3,500$. So if the price falls, then the equity is greater than the required equity, and the maintenance requirement is satisfied. This is summarized in Table 9.6. In fact any equity in excess of 50% of the current price can be used for further short selling. In this case $0.50 \times \$4,000 = \$2,000$, so there is \$1,500 in excess equity. The investor could use this to borrow and sell $\$1,500/0.50 = \$3,000$ worth of stock.

Table 9.6. The Effect of a Price Decrease When a Stock is Sold Short

Credit Balance	Price Per Share	Number of Shares	Market Value	Equity	Required Equity
$7,500	$50	100	$5,000	$2,500	$1,500
$7,500	$40	100	$4,000	$3,500	$1,200

2. The price of stock rises to $60 per share. In this case, the required equity is $0.30 \times \$6,000 = \$1,800$, and the equity is $\$7,500 - \$6,000 = \$1,500$, so the investor pays an extra $300 ($1,800 - \$1,500$) to satisfy the maintenance requirement. If the stock price rises rapidly, then the investor can lose an unlimited amount of money, whereas the investor's profit is limited to the proceeds of the sale less any costs (commissions and interest on the loan of the shares). This makes short selling extremely risky. Table 9.7 illustrates this. Notice that the maintenance fee paid when the price of the stock rises is the difference between the required equity and the equity before the fee is deposited.

Table 9.7. The Effect of a Price Increase When a Stock is Sold Short

Credit Balance	Price Per Share	Number of Shares	Market Value	Equity	Required Equity
$7,500	$50	100	$5,000	$2,500	$1,500
$7,500	$60	100	$6,000	$1,500	$1,800
$7,800	$60	100	$6,000	$1,800	$1,800
$7,800	$70	100	$7,000	$800	$2,100
$9,100	$70	100	$7,000	$2,100	$2,100
$9,100	$80	100	$8,000	$1,100	$2,400
$10,400	$80	100	$8,000	$2,400	$2,400
$10,400	$90	100	$9,000	$1,400	$2,700
$11,700	$90	100	$9,000	$2,700	$2,700
$11,700	$100	100	$10,000	$1,700	$3,000
$13,000	$100	100	$10,000	$3,000	$3,000

In this example, the investor originally deposits $2,500 in the margin account to short sell the stock. When the price reaches $100 per share, the investor has paid an additional $5,500 in maintenance fees.[9] With short selling there is no upper limit to the amount of money one can lose! In this example, once the price of the stock reaches $60 per share, for every $1,000 increase in the market value of the stock the investor must deposit an additional $1,300 to satisfy the maintenance requirement.

Notice from Table 9.7 that somewhere between a share price of $50 and a share price of $60, there must be a price, S, at which the required equity equals the equity, above which a maintenance call is issued. In this case that will occur when

$$7500 - 100S = 0.3 \times 100S,$$

which gives $S = \$57.69$.

A natural question to ask is, "To what level can the stock price rise before a maintenance call is generated?" This question is answered in the following theorem.

Theorem 9.3. *The Short Sale Maintenance Level Theorem.*
If an investor short sells a stock and has an initial credit balance of C_o, and if the maintenance level is m, then the maximum level, V^*, to which the market value of the stock can rise and not generate a maintenance call, is

$$V^* = \frac{1}{1+m} C_o.$$

Proof. If a stock is purchased on margin, if the original market value of the shares purchased is V_o, and if the original equity is E_o, then $C_o = V_o + E_o$. If V is the current market value of the shares, and if E is the current equity, then we have

$$C_o = V + E,$$

as long as the equity has remained at least as large as the required equity. (Why?) At the maximum level V^* we have $E^* = mV^*$ and $C_o = V^* + E^*$, which must be solved for V^*, giving

$$V^* = \frac{C_o}{1+m},$$

or

$$V^* = \frac{V_o + E_o}{1+m}.$$

---

[9] When the price of this stock reaches $100 per share there have been five increases in price, each one generating a maintenance fee. The total of these additional maintenance fees is ($1,800 − $1,500) + ($2,100 − $800) + ($2,400 − $1,100) + ($2,700 − $1,400) + ($3,000 − $1,700) = $5,500.

To show that $(V_o + E_o)/(1 + m)$ is the maximum level, suppose that

$$V > \frac{V_o + E_o}{1 + m}.$$

Then the equity is

$$\begin{aligned}
C_o - V &= (V_o + E_o) - V \\
&< (V_o + E_o) - \frac{V_o + E_o}{1 + m} \\
&= \frac{m}{1 + m}(V_o + E_o).
\end{aligned}$$

But the required equity is

$$mV > \frac{m}{1 + m}(V_o + E_o),$$

so $E = C_o - V < mV$, a maintenance call.   □

Example 9.2. To avoid a maintenance call, what is the maximum level to which the market value of a stock can rise in a margin account containing 100 shares and an initial credit balance of \$7,500 if the maintenance level is 0.3? At what price per share does this occur?

Solution. Here $C_o = 7500$ and $m = 0.3$, so the maximum market value

$$V^* = \frac{10}{13}C_o = \frac{10}{13}\$7,500 = \$5,769.23.$$

The required equity is $0.3 \times \$5,769.23 = \$1,730.77$, and the equity is $\$7,500 - \$5,769.23 = \$1,730.77$. At this level the price per share is $\$5,769.23/100 = \$57.69$, in agreement with our previous result. $\triangle$

9.2 Reading The Wall Street Journal Stock Tables

Section C (Money and Investing) of *The Wall Street Journal* (WSJ) contains tables for the New York Stock Exchange and the NASDAQ. There are explanatory notes for both.

The following stock table is taken from *The Wall Street Journal* of Wednesday, October 11, 2000. The latest trading day, mentioned below, is Tuesday, October 10, and the previous trading day is Monday, October 9.

52 Weeks					Yld		Vol				Net
Hi	Lo	Stock	Sym	Div	%	P/E	100s	Hi	Lo	Close	Chg
49^{58}	26^{38}	McDonalds	MCD	.22	.7	21	19448	30^{38}	29^{94}	29^{94}	-0^{13}

We now explain each of these entries, one by one.

- **52 Weeks Hi:** $49.58 is the highest price of the stock during the preceding 52 weeks including the current week, excluding the latest trading day.
- **52 Weeks Lo:** $26.38 is the lowest price of the stock during the preceding 52 weeks including the current week, excluding the latest trading day.
- **Stock:** McDonalds is the company name.
- **Sym**: MCD is the company ticker symbol. Usually, companies with ticker symbols of three or fewer letters are traded on the NYSE, and those with four or more on the NASDAQ.
- **Div**: $0.22 is the latest annual cash dividend.
- **Yld %**: 0.7% is the dividend expressed as a percentage of the closing price (*Close*), namely $(0.22/29.94) \times 100 = 0.7348$, which is 0.7 rounded to one decimal place.
- **P/E:** 21 is the price to earnings ratio, which is obtained by dividing the closing price (*Close*) by the total earnings per share over the previous four quarters. Thus, the earnings per share is $29.94/21 = 1.4257. There are various ways of interpreting a P/E ratio of 21. One way is: If the price per share stays constant, then it takes 21 years for the total earnings per share to equal the current price per share. A second way is: It costs the stockholder $21 for every $1 the company earns.
- **Vol 100s:** Approximately 1,944,800 McDonalds' shares were traded (daily and unofficial).
- **Hi:** $30.38 is the stock's highest price on the latest trading day.
- **Lo**: $29.94 is the stock's lowest price on the latest trading day.
- **Close:** $29.94 is the stock's closing (that is, last) price on the latest trading day.
- **Net Chg:** $-$0.13 is the change in the closing price of the stock on the latest trading day from the closing price on the previous trading day. Thus, the closing price on the previous trading day is $30.07.

9.3 Problems

Walking

9.1. Use the Internet to find the ticker symbols and closing prices on April 10, 2006 for the following stocks: AT&T, International Paper, and Verizon.

9.2. Use the Internet to find the following information for Microsoft.

(a) What was the first day that Microsoft's stock was traded publicly?

(b) What is the highest price for Microsoft from the date in (a) to the present? (Note that there were several stock splits during this time. What does "highest price" mean in this context?)

9.3. Helen Kendrick buys 10 shares of stock once a month for 4 months. Hugh Kendrick buys $100 worth of the same stock each month at the same time as Helen. Who should have the lower cost per share if the price per share changes over the 4 month period? Verify this if the stock prices are $20, $21, $22, and $23 over the 4 months.

9.4. When is it better to sell a fixed number of shares on a regular basis rather than using dollar cost averaging if one is selling stock?

9.5. Helen Kendrick buys 100 shares on margin. If the margin requirement is 50%, if the maintenance requirement is 25%, and if the stock is selling for $30 a share, then at what price would she receive a maintenance call? Create a table similar to Table 9.4 on p. 155 with stock prices of $30, $25, $20, $15, and $10.

9.6. Referring to Problem 9.5, Hugh Kendrick, believing that the stock price will fall, short sells 100 shares. If the maintenance requirement for short sales is 30%, then at what price would Hugh receive a maintenance call? Create a table similar to Table 9.7 on p. 158 with share prices of $30, $35, $40, $45, and $50.

9.7. Turn to Section C in the latest edition of *The Wall Street Journal*, and find the listing for AT&T. What is the closing price? What is the closing price on the previous trading day? Estimate the earnings per share. How is the earnings calculated? How many shares were traded?

9.8. On which exchange (the NYSE or the NASDAQ) is the stock with ticker symbol CSCO listed? Find the listing in the latest edition of *The Wall Street Journal*. Explain each of the symbols and numbers listed for this stock.

9.9. Let $S(t)$ be the price of XYZ at time t in months. Assume that $S(1) = 10$, $S(2) = 15$, $S(3) = 18$, $S(4) = 20$, and $S(5) = 24$. Wendy and Tom Kendrick each purchase 100 shares of the same stock at time $t = 1$. Show that Tom has a greater profit at time $t = 5$ if he purchases 100 shares at each time, $t = 1, 2, \ldots, 5$, than Wendy, who uses dollar cost averaging each month. Does this contradict what was discussed concerning dollar cost averaging? Explain.

9.10. Hugh Kendrick short sells 100 shares of stock on margin. If the margin requirement is 50%, if the maintenance requirement for short sales is 30%, and if the stock is sold for $60 a share, then how many additional shares can Hugh short sell if the price falls to $55 a share? Create an appropriate table with share prices of $60, $55, $50, $45, and $40.

Questions for Review

- What is a stock split?
- What is dollar cost averaging?
- How do you buy on margin? What are the risks?
- What is short selling? What are the risks?
- What is the Long Sale Maintenance Level Theorem?
- What is the Short Sale Maintenance Level Theorem?

Stock Market Indexes, Pricing, and Risk

Stock market indexes are used to compute an "average" price for groups of stocks. A stock market index attempts to mirror the performance of the group of stocks it represents through the use of one number, the index. Indexes may represent the performance of all stocks in an exchange or a smaller group of stocks, such as an industrial or technological sector of the market. In addition, there are foreign and international indexes.

10.1 Stock Market Indexes

In this section we discuss how Dow Jones, Standard and Poor's, NASDAQ, and Value Line calculate their indexes. These indexes depend on stock price, number of shares issued, stock splits, and dividends. However, these are not independent. For example, if a $50 stock with 100 shares issued splits two-for-one, then after the split there are 200 shares, each worth $25. If a $50 stock pays 10% stock dividend (worth $5 a share), then after the dividend the stock is worth $50/1.1 = $45.45.

Dow Jones Industrial Average

The oldest and most quoted stock market index is the Dow Jones Industrial Average (DJIA) instituted in 1884. Charles Dow created the index and made it public on May 26, 1896. The DJIA originally contained 12 stocks, but stocks were added and deleted until the present number, 30, first appeared in 1928.[1]

The original Dow Jones Average was simply an equal-weighted average of the prices of the stocks that make up the average. However, because of stock

---

[1] There are also Dow Jones indexes for 20 transportation stocks and for 15 utility stocks. These indexes appear in Section C of each issue of *The Wall Street Journal*, unless the NYSE was closed on the previous weekday.

splits, dividends, and the addition and subtraction of stocks from the index, the computation of the DJIA is now more complicated.

As an example, assume that we use the DJIA method to compute an average for stocks S, T, U, and V, whose present share prices are \$15, \$20, \$20, and \$25. Then our index on day 1 is

$$\frac{15 + 20 + 20 + 25}{4} = 20.$$

Now, assume that on day 2 company T has a two-for-one stock split. In this case each shareholder receives two new shares for each old share, and the price of each new share is one-half the price of an old share, \$20/2 = \$10. The DJIA adjusts for the stock split by changing the divisor, 4, so that the DJIA does not change, while the numerator is the sum of the post-split share prices of the stocks. In this example we have

$$\frac{15 + 10 + 20 + 25}{d} = 20,$$

where d is the new divisor. Solving for d gives $d = 3.5$. This new divisor stays in effect until there is another stock split, a stock dividend of 10% or more by one of the companies, or a replacement of an existing company by a new company.

Example 10.1. The table shows the continuing share prices for the companies S, T, U, and V on days 1 through 5, where on day 2, company T had the two-for-one stock split. Complete the table for days 3 through 5 assuming that there are no more stock splits, that there are no dividends, and that no companies are replaced.

Day	Price				Divisor	DJIA
	S	T	U	V		
1	15	20	20	25	4.0	20.00
2	15	10	20	25	3.5	20.00
3	16	11	21	26		
4	15	10	20	25		
5	16	10	19	26		

Solution. Because there are no more stock splits and no dividends, the divisor remains at 3.5 for days 3 though 5. The averages on these days are

$$\frac{16 + 11 + 21 + 26}{3.5} = 21.14 \text{ (Day 3)},$$

$$\frac{15 + 10 + 20 + 25}{3.5} = 20.00 \text{ (Day 4), and}$$

$$\frac{16 + 10 + 19 + 26}{3.5} = 20.29 \text{ (Day 5)}.$$

△

We now discuss the effect of stock splits and dividends on the DJIA and why they may lead to a downward bias (a value smaller than expected) in the DJIA. Suppose that after the stock split (so the divisor is 3.5) company T's stock price increases by 10%, from \$10 to \$11. Then the new average is

$$\frac{15 + 11 + 20 + 25}{3.5} = 20.29.$$

On the other hand, if company U's share price increases by 10%, from \$20 to \$22, while the other share prices do not change, then the new average is

$$\frac{15 + 10 + 22 + 25}{3.5} = 20.57.$$

This suggests that the impact of a stock that splits is less than that of a stock that does not split in computing the average. You are asked to verify this in Problem 10.9.

Now, consider the effect of not adjusting for stock dividends of less than 10%. Suppose that the prices of our four stocks are \$15, \$10, \$20, and \$25, and that $d = 3.5$; the share prices and divisor immediately following the stock split of company T. If company U issues a stock dividend of 5% (\$1 per share), then the resulting stock price is \$20/1.05 = \$19.05, which artificially lowers the average. (Note that the divisor does not change in this case.)

Standard and Poor's 500 Index

Standard and Poor's Corporation publishes several stock indexes that use a different approach than the DJIA. The most followed of these indexes is the Standard and Poor's 500 (S&P 500). There are two essential differences between the S&P 500 and the DJIA. The first is that the S&P 500 includes 500 companies rather than 30, which may better reflect the overall market. The second is that the indexes are computed differently. Because all the S&P indexes are calculated in the same manner, we restrict our discussion to the S&P 500.

The S&P 500 weights all stocks in the index in proportion to the sum of their market capitalizations. The formula is

$$S\&P\,500\,(t) = 10\frac{\sum_{i=1}^{500} S_i(t)N_i(t)}{\sum_{i=1}^{500} S_i(1)N_i(1)}.$$

Here $S_i(1)$ is the price per share of the ith stock on the original date. For the S&P 500 this is the average of the ith stock's prices for the years 1941–1943. The quantity $S_i(t)$ is the price per share of the ith stock at time t, and $N_i(t)$ is the number of shares issued of the ith stock at time t.

Returning to Example 10.1, if we calculate this average at day 1 assuming that 100 shares are issued for each company, then we have

$$\sum_{i=1}^{4} S_i(1)N_i(1) = 15(100) + 20(100) + 20(100) + 25(100) = 8000$$

as the divisor. On day 1 this is also the numerator, so the original index is

$$S\&P\ 500\,(1) = 10(8000/8000) = 10.$$

On day 2, after the stock split we have

$$\sum_{i=1}^{4} S_i(2)N_i(2) = 15(100) + 10(200) + 20(100) + 25(100) = 8000,$$

so that the stock split does not change the divisor. The index following the stock split is

$$S\&P\ 500\,(2) = 10(8000/8000) = 10.$$

With this index all stock splits and stock dividends are accounted for in the numerator. The denominator changes if new stocks are introduced in the index, replacing other stocks. In this case the numerator uses the total market capitalization of the 500 stocks after the replacement, and the denominator is adjusted so that the new average is the same as the old average.

Example 10.2. The table shows the continuing share prices for the companies S, T, U, and V on days 1 through 5, where on day 2, company T had the two-for-one stock split. Complete the table for days 3 through 5, assuming that there are no more stock splits and no dividends.

Day	\multicolumn{4}{c}{Price}	S&P 500
	S	T	U	V	
1	15	20	20	25	10.00
2	15	10	20	25	10.00
3	16	11	21	26	
4	15	10	20	25	
5	16	10	19	26	

Solution. Because there are no more stock splits and no dividends, there are 100 shares issued for companies S, U, and V, and 200 for company T. The indexes on these days are

$$10\left(\frac{16(100) + 11(200) + 21(100) + 26(100)}{8000}\right) = 10.63\ \text{(Day 3)},$$

$$10\left(\frac{15(100) + 10(200) + 20(100) + 25(100)}{8000}\right) = 10.00\ \text{(Day 4)},\ \text{and}$$

$$10\left(\frac{16(100) + 10(200) + 19(100) + 26(100)}{8000}\right) = 10.13\ \text{(Day 5)}.$$

△

NASDAQ Composite Index

Another closely watched index is the NASDAQ Composite Index. It was created in 1971 and currently includes over 2000 stocks. It is computed in the same manner as the S&P 500.

The following table lists the year-end closing values of the DJIA, S&P 500, and NASDAQ Composite Index from 1980 through 2005.

	DJIA	S&P 500	NASDAQ		DJIA	S&P 500	NASDAQ
1980	963.99	135.76	202.34	1993	3754.09	466.45	776.80
1981	875.00	121.57	195.84	1994	3834.44	459.27	751.96
1982	1046.54	140.64	232.41	1995	5117.12	615.93	1052.13
1983	1258.64	164.93	278.60	1996	6448.27	740.74	1291.03
1984	1211.57	167.24	247.10	1997	7908.25	970.43	1570.35
1985	1546.67	211.28	324.90	1998	9181.43	1229.23	2192.69
1986	1895.95	242.17	348.80	1999	11497.12	1469.25	4069.31
1987	1938.83	247.08	330.50	2000	10787.99	1320.28	2470.52
1988	2168.57	277.72	381.40	2001	10021.57	1148.08	1950.40
1989	2753.20	353.40	454.80	2002	8341.63	879.82	1335.51
1990	2633.66	330.22	373.80	2003	10453.92	1111.92	2003.37
1991	3168.83	417.09	586.34	2004	10783.01	1211.92	2175.44
1992	3301.11	435.71	676.95	2005	10717.50	1248.29	2205.32

The IRR, i_{irr}, for the DJIA over these 26 years is

$$i_{\text{irr}} = \left(\frac{10717.50}{963.99} \right)^{1/25} - 1 = 10.114\%.$$

In the same way, the IRR's for the S&P 500 and the NASDAQ Composite Index are 9.280% and 10.026%, respectively. This is consistent with the frequently heard statement that in the long run the stock market gains about 10% a year.

Value Line Index

Value Line published two indexes: the *Value Line Geometric Average*, started in 1961, and the *Value Line Arithmetic Average*, started in 1988. Both indexes use all the (over 1600) stocks covered by *The Value Line Investment Survey*. These stocks include big cap, mid-cap, and small-cap companies, and they are all treated equally. In both indexes stock splits and stock dividends are considered by making the appropriate adjustment to the stock's price.

The Value Line Geometric Average uses a geometric-average of the ratio of stock prices. The base (original) index is 100. There are four steps to this procedure.

1. The closing price of each of the n stocks in *The Value Line Investment Survey* is expressed as a ratio of the preceding day's price, adjusted for stock splits.
2. These ratios are multiplied together.
3. The nth root of this product is calculated, where n is the total number of stocks. This gives the geometric daily ratio.
4. The geometric daily ratio is multiplied by the prior day's index value.

We illustrate this method using stocks S, T, U, and V for a 5-day period with a base index of 100 where T had a two-for-one stock split at the end of day 2.

Day	Stock Price			
	S	T	U	V
1	15	20	20	25
2	15	10	20	25
3	16	11	21	26
4	15	10	20	25
5	16	10	19	26

On day 2 the relative price of stock T is

$$\frac{\text{Price on day 2}}{\text{Adjusted price on day 1}} = \frac{\$10}{\$10} = 1,$$

because the adjusted price at day 1 is $\$20/2 = \$10$, reflecting the two-for-one stock split. All the other relative stock prices are equal to 1. Steps 2 and 3 yield the geometric relative price $(1 \times 1 \times 1 \times 1)^{1/4} = 1$, and Step 4 yields $100 \times 1 = 100$, reflecting the fact that the prices remained the same after adjusting for the stock split of company T.

For day 3, combining Steps 1 through 4, we find that the index is

$$\left(\frac{16}{15} \cdot \frac{11}{10} \cdot \frac{21}{20} \cdot \frac{26}{25}\right)^{1/4} \times 100 = 106.392.$$

In Problem 10.5 you are asked to compute this index for day 5. You might suspect that if all the stock prices rise, then the value of the index increases. Other interesting questions are, "What happens to the index if on some day the stock prices have returned to their original levels? What happens to the index if at some day all stocks have doubled in price?" You are asked to consider these questions in Problems 10.6, 10.7, and 10.8.

We write a formula for this index as follows. On day t let $S_i(t)$ be the closing price of stock i, where $1 \leq i \leq n$, so the previous day's closing price is $S_i(t-1)$. If $VLG(t)$ is the index at the end of day t, so the previous day's index is $VLG(t-1)$, then

$$VLG(t) = VLG(t-1) \left(\frac{S_1(t)}{S_1(t-1)} \frac{S_2(t)}{S_2(t-1)} \cdots \frac{S_n(t)}{S_n(t-1)}\right)^{1/n}. \tag{10.1}$$

The Value Line Arithmetic Average uses an equal-weighted average of the ratio of stock prices. There are three steps to this procedure.

1. The closing price of each of the n stocks in *The Value Line Investment Survey* is expressed as a ratio of the preceding day's price, adjusted for stock splits. This is the same calculation as the Value Line Geometric Average
2. These ratios are added together and divided by n, giving the arithmetic daily ratio.
3. The arithmetic daily ratio is multiplied by the prior day's index value.

With the same notation used in defining $VLG(t)$, the Value Line Arithmetic Average, $VLA(t)$, on day t is

$$VLA(t) = VLA(t-1)\frac{\frac{S_1(t)}{S_1(t-1)} + \frac{S_2(t)}{S_2(t-1)} + \cdots + \frac{S_n(t)}{S_n(t-1)}}{n}. \tag{10.2}$$

These two Value Line Indexes are related by the following theorem.[2]

Theorem 10.1. *The Value Line Theorem.*
The maximum daily ratio attainable by the Value Line Geometric Average is equal to the daily ratio of the Value Line Arithmetic Average. Equality can only occur when every stock in the average has the same percentage price change on a given day.

Proof. On day t, the Geometric Average daily ratio is $VLG(t)/VLG(t-1)$, while the Arithmetic Average daily ratio is $VLA(t)/VLA(t-1)$. Thus, we must show that $VLG(t)/VLG(t-1) \leq VLA(t)/VLA(t-1)$.

From (10.1), we have

$$\frac{VLG(t)}{VLG(t-1)} = \left(\frac{S_1(t)}{S_1(t-1)}\frac{S_2(t)}{S_2(t-1)}\cdots\frac{S_n(t)}{S_n(t-1)}\right)^{1/n},$$

while, from (10.2), we have

$$\frac{VLA(t)}{VLA(t-1)} = \frac{\frac{S_1(t)}{S_1(t-1)} + \frac{S_2(t)}{S_2(t-1)} + \cdots + \frac{S_n(t)}{S_n(t-1)}}{n},$$

so we must show that

$$\left(\frac{S_1(t)}{S_1(t-1)}\frac{S_2(t)}{S_2(t-1)}\cdots\frac{S_n(t)}{S_n(t-1)}\right)^{1/n} \leq \frac{\frac{S_1(t)}{S_1(t-1)} + \frac{S_2(t)}{S_2(t-1)} + \cdots + \frac{S_n(t)}{S_n(t-1)}}{n}.$$

---

[2] Quoted, with minor modifications, from the Value Line article "Comparing the Value Line Averages" at http://www.valueline.com/news/vlv050311.html, accessed on May 23, 2006.

This is just the Arithmetic-Geometric Mean Inequality (see Appendix A.3 on p. 249), with equality if and only if

$$\frac{S_1(t)}{S_1(t-1)} = \frac{S_2(t)}{S_2(t-1)} = \cdots = \frac{S_n(t)}{S_n(t-1)},$$

which occurs when every stock has the same percentage price change.   □

Comparing Indexes

We now compare the DJIA, S&P 500, and the Value Line indexes.

- The DJIA is restricted to 30 of the largest US firms. The S&P 500 takes into account 500 of the largest corporations. Neither index takes into account smaller US firms. The Value Line indexes take into account over 1600 small, medium, and large corporations.
- The DJIA deemphasizes the impact of a stock after a stock split. Stock dividends of 10% or less are not taken into account. From the construction of the index, a stock with a price of $50 has a greater impact than a stock with a price of $10. This is because the numerator of the DJIA is the sum of the current prices of all 30 stocks. Neither the DJIA nor the Value Line indexes take into account the total market value of the stocks because they ignore the number of shares issued. The S&P 500 takes market value into account.
- The use of market value may not always be appropriate for all investors. For example, an individual in a high tax bracket may wish to underweight stocks with high dividend yields (such as public utilities).

10.2 Rates of Return for Stocks and Stock Indexes

We now consider rates of returns for stocks and stock indexes. For a given stock, the rate of return[3] during a given time period is defined as the sum of the change in the price of the stock during that period plus any cash dividends received during that period, divided by the price of the stock at the beginning of the period. So if $S(0)$ is the price of the stock at the beginning of the period, if $S(1)$ is the price of the stock at the end of the period, and if D is the cash dividend, then the rate of return R is

$$R = \frac{S(1) - S(0) + D}{S(0)}.$$

From this we see that

$$1 + R = \frac{S(1) + D}{S(0)},$$

---

[3] This rate of return should not be confused with the internal rate of return.

which is positive because $S(0) > 0$, $S(1) > 0$, and $D \geq 0$. This property will be used later.

Example 10.3. A stock is trading at \$50 a share on 12/31/2004. The following is a list of the stock prices and the dividends paid from 1/1/2005 through 12/31/2005 on the last day of each quarter. What are the rates of return for each of the quarters?

Date	3/31/2005	6/30/2005	9/30/2005	12/31/2005
Price	\$52.00	\$54.00	\$53.00	\$56.00
Dividend	\$0.50	\$0.50	\$0.50	\$0.50

Solution. The rates of return for the four quarters are

$$\frac{(52.00 - 50.00) + 0.50}{50.00} = 0.05 = 5\%,$$

$$\frac{(54.00 - 52.00) + 0.50}{52.00} = 0.0481 = 4.81\%,$$

$$\frac{(53.00 - 54.00) + 0.50}{54.00} = -0.0093 = -0.93\%,$$

and

$$\frac{(56.00 - 53.00) + 0.50}{53.00} = 0.066 = 6.6\%.$$

$\triangle$

If R_1, R_2, R_3, and R_4 are the quarterly rates of return (expressed as decimals), then how can we quantify the average quarterly rate of return, R? This depends on whether we use simple or compound interest.

If we use simple interest, then the year-end value of a \$1 investment is $1 + R_1 + R_2 + R_3 + R_4$, and we want this to be equal to $1 + 4R$, so

$$1 + 4R = 1 + R_1 + R_2 + R_3 + R_4.$$

Solving this for R, gives the ARITHMETIC AVERAGE QUARTERLY RETURN,[4]

$$R_A = \frac{R_1 + R_2 + R_3 + R_4}{4}.$$

If we use compound interest, then the year-end value of a \$1 investment is $(1 + R_1)(1 + R_2)(1 + R_3)(1 + R_4)$, and we want this to equal $(1 + R)^4$, so

$$(1 + R)^4 = (1 + R_1)(1 + R_2)(1 + R_3)(1 + R_4).$$

Solving this for R, gives the GEOMETRIC AVERAGE QUARTERLY RETURN,[5]

---

[4] Strictly speaking this should be called a *rate of return*, but for the sake of brevity, in this section we use the shorter expression *return*.

[5] Note that, although the Arithmetic Average Quarterly Return of R_1, R_2, R_3, R_4 is their arithmetic mean, the Geometric Average Quarterly Return of R_1, R_2, R_3, R_4 is not their geometric mean.

$$R_G = ((1 + R_1)(1 + R_2)(1 + R_3)(1 + R_4))^{1/4} - 1.$$

We have already shown that each of the terms in parentheses is positive, so R_G is always defined.

Example 10.4. For the stock in Example 10.3, calculate the arithmetic and geometric average quarterly returns.

Solution. The arithmetic average quarterly return is

$$\frac{0.05 + 0.0481 + (-0.0093) + 0.066}{4} = 0.0387 = 3.87\%,$$

while the geometric average quarterly return is

$$((1.05)(1.0481)(0.9907)(1.066))^{1/4} - 1 = 0.0383 = 3.83\%.$$

$\triangle$

Note that the geometric average quarterly return is less than the arithmetic average quarterly return. This is generally true, as can be seen in the following theorem.

Theorem 10.2. *The Average Quarterly Return Theorem.*

$$R_G \leq R_A.$$

Proof. Using the Arithmetic-Geometric Mean Inequality (see Appendix A.3 on p. 249), which states that if $a_1, a_2, \ldots, a_n$ are non-negative and not all zero, then

$$(a_1 a_2 \cdots a_n)^{1/n} \leq \frac{a_1 + a_2 + \cdots + a_n}{n},$$

with equality if and only if $a_1 = a_2 = \cdots = a_n$. With $n = 4$ and $a_i = 1 + R_i > 0$ for $i = 1, \ldots, 4$, this inequality becomes

$$((1 + R_1)(1 + R_2) \cdots (1 + R_4))^{1/4} \leq \frac{(1 + R_1) + (1 + R_2) + \cdots + (1 + R_4)}{4},$$

with equality if and only if $R_1 = R_2 = R_3 = R_4$. This shows that the geometric average quarterly return is less than the arithmetic average quarterly return unless the quarterly returns are equal, in which case the averages are the same. $\square$

There are several things to note about these definitions of return.

- It is assumed that dividends are paid on the last day of each quarter. If this is not the case, then the calculation is done daily.
- It is assumed that dividends are not reinvested in the stock.
- We cannot calculate this form of return if we do not know the amounts and timing of the dividend payments. For example, we cannot calculate the return on the S&P 500 from a listing of the year-end values of that index.

10.3 Pricing and Risk

As defined in the previous section, the return on an investment in common stock consists of two components: cash dividends and the rise or fall of the stock price, capital gains or losses. The price that an investor is willing to pay for a share of common stock is based upon the investor's expectations regarding dividends and the future price of the stock.

Let $S(0)$ be the current price per share of the stock, D_t be the expected dividend at time t in quarters, T be the length of the holding period in quarters, and k be the quarterly required rate of return (expressed as a decimal).

Initially, we assume that dividends are paid quarterly and that the price of the stock is to be determined immediately following a dividend payment. We also ignore certain aspects of the trading mechanism.

Theoretically, the length of the holding period could be infinite. Common stocks do not mature, unlike bonds. As long as the company continues to exist, then so does the stock. The investor expects to receive dividends in the amount of D_1 at the end of the first quarter, D_2 at the end of the second quarter, and so on. In return, the investor is willing to pay $S(0)$ for the stock today. These cash flows are represented by the diagram in Fig. 10.1.

Fig. 10.1. Time diagram of a typical stock held forever

The price that the investor is willing to pay for the stock today is given by the present value of the expected future dividends discounted at the appropriate required rate of return.[6] Assuming that the appropriate required rate of return per quarter is k, the general formula for the price of the stock today is

$$S(0) = \frac{D_1}{(1+k)^1} + \frac{D_2}{(1+k)^2} + \cdots,$$

or

$$S(0) = \sum_{t=1}^{\infty} \frac{D_t}{(1+k)^t}. \tag{10.3}$$

In practice, however, the length of the holding period is not infinite. The investor expects to receive dividends in the amount of D_1 at the end of the first quarter, D_2 at the end of the second quarter, and so on. In addition, the

---

[6] We discuss the required rate of return on p. 177 and the relationship between the required rate of return and the expected rate of return on p. 184.

investor expects to be able to sell the stock for a price of $S(T)$ at the end of T quarters. In return, the investor is willing to pay $S(0)$ for the stock today. These cash flows are represented by the diagram in Fig. 10.2.

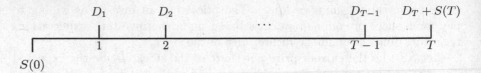

Fig. 10.2. Time diagram of a typical stock

Again, the price that the investor is willing to pay for the stock today is given by the present value of the expected future cash flows discounted at the appropriate required rate of return. Therefore, an alternative formula for the price of the stock today is given by

$$S(0) = \frac{D_1}{(1+k)^1} + \frac{D_2}{(1+k)^2} + \cdots + \frac{D_{T-1}}{(1+k)^{T-1}} + \frac{D_T + S(T)}{(1+k)^T},$$

or

$$S(0) = \sum_{t=1}^{T} \frac{D_t}{(1+k)^t} + \frac{S(T)}{(1+k)^T}. \tag{10.4}$$

Thus, we have the following theorem.

Theorem 10.3. *The Common Stock Theorem.*
If the price $S(0)$ of a share of common stock is given by (10.3), then the price may be written as (10.4).

The converse of this theorem is not necessarily true (see Problem 10.13).

Example 10.5. Helen Kendrick expects to receive dividends in the amount of $1.50 at the end of each of the next four quarters and $1.75 at the end of each of the following four quarters. Furthermore, she expects to be able to sell the stock for $94.50 at the end of the second year. If the required rate of return per quarter is 0.025, then what price should she pay for the stock?

Solution. Here $T = 8$, $D_1 = D_2 = D_3 = D_4 = 1.5, D_5 = D_6 = D_7 = D_8 = 1.75$, $k = 0.025$, and $S(T) = 94.5$. Substituting these into (10.4) gives

$$S(0) = \sum_{t=1}^{4} \frac{1.5}{(1.025)^t} + \sum_{t=5}^{8} \frac{1.75}{(1.025)^t} + \frac{94.5}{(1.025)^8} = 89.168,$$

so Helen should pay $89.17 for the stock. $\triangle$

Dividends tend to be relatively stable over the short term; however, there is usually variability in a perpetual stream of dividends. This creates difficulties in applying the general pricing formula. However, it is often assumed that the dividends paid by a company to its shareholders will grow indefinitely at a fixed rate. If we let g be the quarterly rate of dividend growth (expressed as a decimal), then

$$D_1 = D_0(1+g),$$
$$D_2 = D_1(1+g) = D_0(1+g)^2,$$

and so on.

According to the general pricing formula (10.3),

$$S(0) = \sum_{t=1}^{\infty} \frac{D_0 (1+g)^t}{(1+k)^t}. \tag{10.5}$$

This leads to the following theorem.

Theorem 10.4. *The Constant Growth Theorem.*
If the dividend grows according to $D_n = D_0(1+g)^n$ and if $k > g$, then

$$S(0) = D_0 \frac{1+g}{k-g}.$$

Proof. We can rewrite (10.5) as

$$S(0) = \sum_{t=1}^{\infty} D_0 \left(\frac{1+g}{1+k}\right)^t = D_0 \left(\frac{1+g}{1+k}\right) \sum_{t=0}^{\infty} \left(\frac{1+g}{1+k}\right)^t,$$

which is a geometric series. Because $k > g$, we have $(1+g)/(1+k) < 1$, which means that the series converges. Thus,

$$S(0) = D_0 \left(\frac{1+g}{1+k}\right) \frac{1}{1 - \frac{1+g}{1+k}} = D_0 \frac{1+g}{k-g}.$$

$\square$

Example 10.6. A company just paid a dividend of \$2.25, and future dividends are expected to increase at a rate of 1% per quarter. If the quarterly rate of return on the company's stock is 2%, then what price should the investor pay for the stock?

Solution. Here $D_0 = 2.25$, $g = 0.01$, and $k = 0.02$, so $k > g$ and thus

$$S(0) = D_0 \frac{1+g}{k-g} = 2.25 \frac{1.01}{0.01} = 227.25.$$

The investor should pay \$227.25. $\triangle$

When discussing how to price a stock, we talk about using an appropriate risk-adjusted rate to discount the dividends that are expected to be paid on the stock in the future. This is the required rate of return mentioned on p. 175.

However, two crucial questions remain unanswered at this point. How do we measure the risk of a stock? How do we use that measure of risk to compute the appropriate discount rate? These questions are now addressed.

The risk of a stock is associated with uncertainty about the returns on the stock. If the return is certain, then there is no risk. As the uncertainty of the returns increases, then so does risk. Risk can be estimated using any one of a number of measures. These include the range, the mean absolute deviation, the probability of a negative return, the semivariance, and the standard deviation of the returns on the stock.

We assume that the returns, R_s, on a stock have S possible values, so $1 \leq s \leq S$, and that the probability of R_s is P_s. If R is the random variable that gives the return on the stock, then the possible values of R are $R_1, R_2, \ldots, R_S$, and the expected return on the stock is $E(R) = \sum_{s=1}^{S} P_s R_s$.

The **range** is the difference between the largest and smallest returns. If the largest return is $\max_{s \in S} R_s$ and the smallest return is $\min_{s \in S} R_s$, then the range is

$$\max_{s \in S} R_s - \min_{s \in S} R_s.$$

This is simple to compute; however, it does not consider returns between the extremes and it does not consider the likelihoods of the possible returns.

The **mean absolute deviation** is the expected absolute difference between the return on the stock and its expected return, $E(R)$, and is given by

$$MAD = \sum_{s=1}^{S} P_s |R_s - E(R)|.$$

This may be a reasonable measure, but it is statistically difficult to use.

The **probability of a negative return** is the sum of the probabilities of all negative returns and is given by[7]

$$\sum_{s=1}^{S} P_s 1_{R_s < 0}.$$

The weakness of this measure is that it does not consider all aspects of risk. For example, neither non-negative returns nor the magnitudes of negative returns are considered.

Semivariance is a statistical measure of the variability of returns below the expected return and is given by

$$\sum_{s=1}^{S} P_s (R_s - E(R))^2 1_{R_s < E(R)}.$$

---

[7] If X is a random variable, and $(X \subset A)$ is an event, then the **indicator variable** $1_{X \subset A}$ is defined by

$$1_{X \subset A} = \begin{cases} 1 \text{ if } X \subset A, \\ 0 \text{ if } X \not\subset A. \end{cases}$$

As with the probability of a negative return, the semivariance does not consider the uncertainty of returns greater than the expected return.

The **standard deviation** of R is one of the most commonly used measures of risk. It is a measure of the dispersion both above and below the expected return and is given by

$$\sigma = \sqrt{\sum_{s=1}^{S} P_s(R_s - E(R))^2} \ .$$

The quantity σ^2 is the **variance** of R.

Example 10.7. Suppose that an investment has three possible annual rates of return: a return of 0.10 with probability 0.50, a return of 0.15 with probability 0.30, and a return of -0.05 (a loss of 5%) with probability 0.20. Thus, if we let R be the annual rate of return, then $P(R = 0.10) = 0.50$, $P(R = 0.15) = 0.30$, and $P(R = -0.05) = 0.20$. What is the range, expected return, MAD, probability of a negative return, semivariance, and standard deviation?

Solution. The range is $0.15 - (-0.05) = 0.20$.

The expected return is $E(R) = 0.5(0.10) + 0.3(0.15) + 0.2(-0.05) = 0.085$.
The MAD is $0.5|0.10 - 0.085| + 0.3|0.15 - 0.085| + 0.2|-0.05 - 0.085| = 0.054$.
The probability of a negative return is 0.2.
The semivariance is $0.2(-0.05 - 0.085)^2 = 0.0036$.
The standard deviation is

$$\sigma = \sqrt{0.5(0.10 - 0.085)^2 + 0.3(0.15 - 0.085)^2 + 0.2(-0.05 - 0.085)^2}$$
$$= 0.0709.$$

$\triangle$

These measures of risk consider two types of risk—systematic risk and unsystematic risk. Systematic risk is risk that is common to all risky stocks. Unsystematic risk is unique to each company. We discuss each of these in turn.

Systematic risk includes the impact of inflation, uncertainties about long-run aggregate economic growth, changes in investors' attitudes toward risk, and changes in interest rates.

- Inflation affects stock prices through corporate profits. If inflation results in higher borrowing rates, then expected future corporate profits are likely to decrease. On the other hand, if inflation results in higher prices, then expected future corporate profits are likely to increase. The effect is similar for all stocks.

- Uncertainties about long-run aggregate economic growth also affect stock prices through corporate profits. Corporate profits of all firms are dependent upon the long-run health and growth of the aggregate economy. Factors that affect aggregate output include population growth, labor productivity, political uncertainty, tax policy, technology, etc.
- Changes in investors' attitudes toward risk have a more direct affect on stock prices. If the average investor suddenly becomes more risk-averse, then the required rate of return increases. To provide this larger return, stock prices have to fall. The opposite occurs if the average investor suddenly becomes less risk-averse. Again, the effect is similar for all stocks.
- Interest rate changes also affect the prices of all stocks. Increases in interest rates result in increases in interest expense and decreases in corporate profits. Changes in interest rates also affect investors' attitudes toward risk that in turn affect stock prices.

Systematic risk cannot be eliminated through diversification. For this reason it is often referred to as market or nondiversifiable risk.

Unsystematic risk is unique to each company. This type of risk includes lawsuits, strikes, competition, and changes in consumer base.

- If one company sues another for breach of contract, then the market value of the winning company increases, and the market value of the losing company decreases. An appropriate position in both stocks eliminates the need for any concern about the outcome of the lawsuit.
- If a company suffers a labor strike, then that company's profit decreases. However, if the lost sales are picked up by a competitor, then that company's profit increases. An appropriate position in both stocks eliminates the need for any concern about the outcome of the strike.
- If one company has a patent on a major product line that may be counteracted through research and development at another company, then the profits of the company with the patent are likely to decrease with the loss of its effective patent position, and the profits of the competitor are likely to increase. Once again, an appropriate position in both stocks eliminates the need for any concern.
- If the companies in one region lose customers to companies in another region as a result of migration, then the profits of companies in the first region decrease, and the profits of companies in the second region increase. Yet again, an appropriate position in the stocks of companies in both regions eliminates the need for any concern.

Unsystematic risk can be eliminated by investing in a well diversified portfolio. Losses in one stock are offset by gains in another stock. For this reason, unsystematic risk is often referred to as diversifiable risk.

Because unsystematic risk can be eliminated through proper diversification, investors are not rewarded or compensated for that type of risk; investors are only rewarded for systematic risk. In order to determine how much of a stock's return is not diversifiable, we need a measure of the relationship of the returns on that stock with the returns on other stocks.

The correlation coefficient is a commonly used statistical measure of the relationship between two random variables. We assume that the returns on stocks i and j, namely, R_{is} and R_{js}, have S possible values, so $1 \leq s \leq S$, that the probabilities of R_{is} and R_{js} are the same, P_s, that the expected returns on stocks i and j are $E(R_i)$ and $E(R_j)$ respectively, and that the standard deviation of the return on stocks i and j are σ_i and σ_j respectively. Then the correlation coefficient between the returns on stock i and the returns on stock j is given by

$$\rho_{ij} = \frac{\sum\limits_{s=1}^{S} P_s(R_{is} - E(R_i))(R_{js} - E(R_j))}{\sigma_i \sigma_j}.$$

Example 10.8. In addition to the information given for the annual rate of return R of the investment in Example 10.7, suppose that there is a second investment for which the annual rate of return U is correlated with R as follows: $P(R = 0.10, U = 0.15) = 0.50$, $P(R = 0.15, U = 0.20) = 0.30$, and $P(R = -0.05, U = -0.10) = 0.20$. Find ρ_{RU}, the correlation coefficient between R and U.

Solution. The expected return on the second stock is $E(U) = 0.5(0.15) + 0.3(0.20) + 0.2(-0.10) = 0.115$, and the standard deviation is

$$\sigma = \sqrt{0.5(0.15 - 0.115)^2 + 0.3(0.20 - 0.115)^2 + 0.2(-0.10 - 0.115)^2}$$
$$= 0.1097.$$

Therefore, the correlation coefficient is

$$\rho_{RU} = \frac{1}{0.0709(0.1097)}[0.5(0.10 - 0.085)(0.15 - 0.115)$$
$$+ 0.3(0.15 - 0.085)(0.20 - 0.115)$$
$$+ 0.2(-0.05 - 0.085)(-0.10 - 0.115)]$$
$$= 0.993. \quad \triangle$$

If the correlation coefficient between the returns on two stocks is -1, then risk can be eliminated through proper diversification. If the correlation coefficient between the returns on two stocks is 1, then risk cannot be eliminated through diversification.

However, it is not sufficient to consider the relationship between just two stocks; we must consider the relationships among all stocks. Then one appropriate measure of the risk of a stock is the amount of nondiversifiable risk that the stock contributes to that of the market portfolio. For stocks issued by domestic companies, an index such as the S&P 500 or the NYSE Composite Index[8] is used as a proxy for the market portfolio. This measure of risk is given by $\sigma_i \rho_{iM}$, where σ_i is the standard deviation of the returns on the stock, and ρ_{iM} is the correlation coefficient between the returns on the stock and the returns on the market.

A more commonly used measure of the risk of a stock is the amount of nondiversifiable risk inherent in the stock relative to that of the market portfolio. This measure of the relative risk of a stock is known as the stock's BETA and is given by $\beta_i = (\sigma_i/\sigma_M)\rho_{iM}$, where σ_i is the standard deviation of the returns on the stock, σ_M is the standard deviation of the returns on the market, and ρ_{iM} is the correlation coefficient between the returns on the stock and the returns on the market. For example, if R_M represents the returns on the S&P 500, then ρ_{iM} is the correlation coefficient between R_M and the returns on the given stock.

The ratio of σ_i to σ_M measures how volatile the stock is in relation to the volatility of the reference portfolio, and the correlation coefficient determines how much of that relative volatility should be counted. If the stock is perfectly correlated with the market, then all of the relative volatility counts. If the correlation is zero, then none of the relative volatility counts.

In practice, investors do not know σ_i, σ_M, or ρ_{iM}. Therefore, these parameters must be estimated. What is needed are ex ante[9] estimates of σ_i, σ_M, and ρ_{iM}.

However, the use of ex ante values requires that all possible future returns and the corresponding probabilities be identified—an extremely difficult, if not impossible, task. Thus, ex post[10] or historical estimates of σ_i, σ_M, and ρ_{iM} are often used as estimates of the ex ante values of σ_i, σ_M, and ρ_{iM} respectively.

A regression line is a line that describes the relationship between an explanatory variable and a response variable. The characteristic line for a stock is the best-fit regression line that describes the relationship between the returns on the stock and the returns on the market. The slope of this line is given by $b_i = (s_i/s_M)r_{iM}$, where s_i and s_M are estimates of σ_i and σ_M respectively, and r_{iM} is an estimate of ρ_{iM} (see Appendix B.7 on p. 280). Thus, b_i is the estimate of β_i. In other words, the estimate of a stock's β is obtained by regressing the stock's historical returns against the returns on the market.

Although this sounds simple, there are several questions that remain. Should β_i be estimated over a time period that consists of the past month, the past year, the past five years, the past 20 years? Should β_i be estimated using daily, weekly, monthly, quarterly, or annual returns? Which index should

---

[8] The NYSE Composite Index includes all common stocks traded on the NYSE. Details of its calculation can be found at www.nyse.com/pdfs/methodology_nya.pdf.

[9] Ex ante means *before the fact*.

[10] Ex post means *after the fact*.

be used as a proxy for the market portfolio? Unfortunately these questions remain unanswered. Therefore, it is important to emphasize the fact that b_i is only an estimate of β_i and that the value of that estimate is determined by the particular data that was used in the regression. In other words, the use of different data may result in a different estimate. However, in spite of the difficulties associated with estimating β_i, it is accepted as an appropriate measure of risk.

Example 10.9. Closing prices and weekly returns for Amazon.com and the S&P 500 from December 1999 to June 2000 are given in following table.[11] Use a spreadsheet to estimate Amazon.com's β.

Week of	Amazon.com Close	Amazon.com Return	S&P 500 Close	S&P 500 Return
26-Jun-00	36.3125	0.071956	1454.60	0.009102
19-Jun-00	33.8750	−0.263587	1441.48	−0.015692
12-Jun-00	46.0000	−0.118563	1464.46	0.005155
5-Jun-00	52.1875	−0.098272	1456.95	−0.013748
29-May-00	57.8750	0.244624	1477.26	0.072016
22-May-00	46.5000	−0.116390	1378.02	−0.020562
15-May-00	52.6250	−0.020930	1406.95	−0.009860
8-May-00	53.7500	−0.081197	1420.96	−0.008146
1-May-00	58.5000	0.060023	1432.63	−0.013632
24-Apr-00	55.1875	0.053699	1452.43	0.012471
17-Apr-00	52.3750	0.117333	1434.54	0.057484
10-Apr-00	46.8750	−0.306198	1356.56	−0.105378
3-Apr-00	67.5625	0.008396	1516.35	0.011858
27-Mar-00	67.0000	−0.078246	1498.58	−0.018907
20-Mar-00	72.6875	0.121504	1527.46	0.043012
13-Mar-00	64.8125	−0.030841	1464.47	0.049747
6-Mar-00	66.8750	0.070000	1395.07	−0.010006
28-Feb-00	62.5000	−0.095841	1409.17	0.056856
21-Feb-00	69.1250	0.067568	1333.36	−0.009457
14-Feb-00	64.7500	−0.150123	1346.09	−0.029579
7-Feb-00	76.1875	−0.030231	1387.12	−0.026152
31-Jan-00	78.5625	0.273556	1424.37	0.047208
24-Jan-00	61.6875	−0.006042	1360.16	−0.056336
17-Jan-00	62.0625	−0.034047	1441.36	−0.016237
10-Jan-00	64.2500	−0.076370	1465.15	0.016428
3-Jan-00	69.5625	−0.086207	1441.47	−0.018908
27-Dec-99	76.1250	−0.154167	1469.25	0.007481
20-Dec-00	90.0000		1458.34	

---

[11] Data obtained from http://finance.yahoo.com/.

Solution. Microsoft® Excel was used to regress the returns for Amazon.com on the returns for the S&P 500. The slope of the regression line is 2.214659. Therefore, an estimate of Amazon.com's β is 2.21. $\triangle$

The price of a stock is determined by discounting all expected future dividends at the appropriate required rate of return. A model known as the Capital Asset Pricing Model (CAPM) specifies the equilibrium relationship between expected returns and risk for stocks. Equilibrium is defined to be a state in which the expected rates of return are equal to the required rates of return. According to this model, investors should be compensated for the time value of money and for nondiversifiable risk.

The line that relates the expected return on a portfolio, $E(R)$, where R is the random variable that gives the return on the portfolio, to the standard deviation of the returns on the portfolio, σ, is called the Capital Market Line. The equation of this line in the $(\sigma, E(R))$ plane is given by

$$E(R) = R_{rf} + \frac{E(R_M) - R_{rf}}{\sigma_M}\sigma,$$

where R_{rf} is the risk-free rate of return,[12] $E(R_M)$ is the expected return on the market, and σ_M is the standard deviation of the returns on the market. The Capital Market Line is the efficient boundary for portfolios, so, under the assumption that the market portfolio is efficient, all portfolios, as plotted in the $(\sigma, E(R))$ plane, must lie on or below this line

Now, we show how the expected rate of return on an individual asset relates to its individual risk.

Consider a portfolio consisting of a proportion w invested in asset i and a proportion $1 - w$ invested in the market portfolio. The expected rate of return on this portfolio is

$$E(R_{i,w}) = wE(R_i) + (1 - w)E(R_M),$$

where $R_{i,w}$ is the random variable that gives the return on the portfolio and R_i is the random variable that gives the return on asset i.[13] This portfolio has standard deviation[14]

$$\sigma_{i,w} = \sqrt{w^2\sigma_i^2 + 2w(1 - w)\sigma_{iM} + (1 - w)^2\sigma_M^2},$$

---

[12] The risk-free rate of return is usually the rate of return on a United States Treasury bill.

[13] Thus, $R_{i,0} = R_M$ and $R_{i,1} = R_i$. Do not confuse the R_i used here, which is the return on asset i, with the R_s used on p. 178, which is one of the possible values of R, the return on a stock.

[14] Thus, $\sigma_{i,0} = \sigma_M$ and $\sigma_{i,1} = \sigma_i$.

where σ_i is the standard deviation of R_i, σ_M is the standard deviation of R_M, and σ_{iM} is the covariance of R_i and R_M.

As we vary w, we obtain different points $(\sigma_{i,w}, E(R_{i,w}))$ in the $(\sigma, E(R))$ plane. When $w = 0$ we are fully invested in the market, so $R_{i,0} = R_M$ and $\sigma_{i,0} = \sigma_M$, and we intersect the Capital Market Line at $(\sigma_M, E(R_M))$.

By the definition of the Capital Market Line, points $(\sigma_{i,w}, E(R_{i,w}))$ can never cross this line. Hence the curve $(\sigma_{i,w}, E(R_{i,w}))$, $-\infty < w < \infty$, is tangent to the Capital Market Line at $(\sigma_M, E(R_M))$. Thus, the slope of this curve at $(\sigma_M, E(R_M))$ must be equal to the slope of the Capital Market Line at $(\sigma_0, E(R_0))$.

The slope of the Capital Market Line is

$$\text{Slope} = \frac{E(R_M) - R_{rf}}{\sigma_M},$$

while the slope of our curve at w is

$$\frac{dE(R_{i,w})}{d\sigma_{i,w}} = \frac{\frac{dE(R_{i,w})}{dw}}{\frac{d\sigma_{i,w}}{dw}} = \frac{E(R_i) - E(R_M)}{(w\sigma_i^2 + (1 - 2w)\sigma_{iM} - (1 - w)\sigma_M^2)/\sigma_{i,w}}.$$

Thus, when $w = 0$ we must have

$$\frac{E(R_M) - R_{rf}}{\sigma_M} = \frac{E(R_i) - E(R_M)}{(\sigma_{iM} - \sigma_M^2)/\sigma_M}.$$

If we rearrange the terms, then we have

$$E(R_i) = R_{rf} + \beta_i(E(R_M) - R_{rf}), \tag{10.6}$$

where

$$\beta_i = \frac{\sigma_i}{\sigma_M}\rho_{iM}.$$

We interpret (10.6) as follows. The expected return (which is the same as the required rate of return in equilibrium) is equal to the risk-free rate of return plus a risk premium. The risk premium is proportional to β_i and the market risk premium, the difference between the expected return on the market and the risk-free rate.

Example 10.10. Suppose that the risk-free rate is $R_{rf} = 0.06$, the expected return on the market is $E(R_M) = 0.15$, and beta is $\beta_i = 1.2$. What is the expected rate of return?

Solution From (10.6), the expected rate of return is $E(R_i) = 0.06 + 1.2(0.15 - 0.06) = 0.168$. $\triangle$

10.4 Portfolio of Stocks

As with bonds, prudent investors invest in portfolios of stocks rather than in individual stocks. The primary advantage of this strategy is risk reduction, a topic discussed in the previous section. In particular, we saw that investors are not compensated for diversifiable risk. A portfolio of stocks consists of investments in several underlying stocks. Therefore, the value of the portfolio is a function of the values of the underlying stocks, and the risk of the portfolio is a function of the risks of the underlying stocks.

The value of a portfolio is simply the sum of the values of the underlying stocks. If the number of different stocks in a portfolio is given by n, the current price of stock i is given by S_i, and the number of shares of stock i in the portfolio is given by N_i, then the value of the portfolio is given by

$$\text{Value of portfolio} = \sum_{i=1}^{n} N_i S_i.$$

Example 10.11. An investor owns 150 shares of stock X and 200 shares of stock Y. If the current prices of stocks X and Y are \$50 and \$35 respectively, then what is the value of the portfolio?

Solution. The value of the portfolio is $150(\$50) + 200(\$35) = \$14,500.$ △

As with individual stocks, one of the most commonly used measures of the volatility of the returns on a portfolio of stocks is the standard deviation. Similarly, a commonly used measure of the risk of a portfolio of stocks is β.

The variance of a portfolio is not the weighted average of the variances of the underlying stocks. Nor is the standard deviation of the returns on a portfolio the weighted average of the standard deviations of the returns on the underlying stocks. Recall that the returns on a stock are often correlated with the returns on other stocks. These correlations must be considered in the computation of the standard deviation.

The fraction of the portfolio that is invested in stock i is given by

$$w_i = \frac{N_i S_i}{\sum\limits_{j=1}^{n} N_j S_j}.$$

If P is the random variable that gives the return on the portfolio, if σ_i and σ_j are the standard deviations of the returns on stocks i and j respectively, and if ρ_{ij} is the correlation coefficient between the returns on stock i and the returns on stock j, then the standard deviation of the returns on the portfolio is given by

$$\sigma_P = \sqrt{\sum_{i=1}^{n} w_i^2 \sigma_i^2 + 2 \sum_{i=1}^{n-1} \sum_{j=i+1}^{n} w_i w_j \sigma_i \sigma_j \rho_{ij}} \ .$$

Example 10.12. A portfolio consists of three stocks: Stock 1, Stock 2, and Stock 3. The following table shows the fraction of the portfolio that is invested in each stock and the standard deviation of the returns on each stock.

Stock	Fraction of Portfolio	σ
1	0.25	0.05
2	0.40	0.07
3	0.35	0.02

The correlation coefficient between the returns on Stock 1 and the returns on Stock 2 is -0.80, the correlation coefficient between the returns on Stock 1 and the returns on Stock 3 is -0.40, and the correlation coefficient between the returns on Stock 2 and the returns on Stock 3 is 0.10. What is the standard deviation of the returns on the portfolio?

Solution. We find that

$$\sigma_P = [(0.25)^2(0.05)^2 + (0.40)^2(0.07)^2 + (0.35)^2(0.02)^2$$
$$+ \; 2(0.25)(0.40)(0.05)(0.07)(-0.80)$$
$$+ \; 2(0.25)(0.35)(0.05)(0.02)(-0.40) + 2(0.40)(0.35)(0.07)(0.02)(0.10)]^{1/2}$$
$$= 0.02.$$

$\triangle$

There are two equivalent ways to compute the beta of a portfolio of stocks.

1. If the standard deviation of the returns on the portfolio is σ_P, if the standard deviation of the returns on the market is σ_M, and if the correlation coefficient between the returns on the portfolio and the returns on the market is ρ_{pM}, then the beta of the portfolio is

$$\beta_P = \frac{\sigma_P}{\sigma_M}\rho_{PM}.$$

2. The beta of a portfolio of stocks is also the weighted average of the betas of the stocks that comprise the portfolio. If the number of stocks in the portfolio is n, if the beta of stock i is β_i, and if the weight of stock i is w_i, then the beta of the portfolio is

$$\beta_P = \sum_{i=1}^{n} w_i\beta_i.$$

That these are equivalent is seen by showing[15] they are both equal to σ_{PM}/σ_M^2 as follows:

$$\beta_P = \frac{\sigma_P}{\sigma_M}\rho_{PM} = \frac{\sigma_P}{\sigma_M}\frac{\sigma_{PM}}{\sigma_P\sigma_M} = \frac{\sigma_{PM}}{\sigma_M^2},$$

---

[15] Recall that the covariance of X and Y satisfies $\sigma_{XY} = \rho_{XY}\sigma_X\sigma_Y$.

while

$$\beta_P = \sum_{i=1}^{n} w_i \beta_i = \sum_{i=1}^{n} w_i \frac{\sigma_i}{\sigma_M} \rho_{iM} = \sum_{i=1}^{n} w_i \frac{\sigma_{iM}}{\sigma_M^2} = \frac{\sigma_{PM}}{\sigma_M^2}.$$

Example 10.13. A portfolio consists of 25% of a stock with a beta of 1.10, 30% of a stock with a beta of 0.95 and 45% of a stock with a beta of 1.30. What is the beta of the portfolio?

Solution. The beta of the portfolio is $0.25(1.10) + 0.30(0.95) + 0.45(1.30) = 1.145$. $\triangle$

10.5 Problems

Walking

10.1. The following table contains data on three stocks: S, T, and U.

	12/31/98	3/31/99	6/30/99	9/30/99	12/31/99
Price S	25	27	28	28	28
T	30	32	33	35	33
U	40	42	43	45	43
Shares S	100	200	200	200	200
Issued T	500	500	500	500	500
U	300	450	450	550	550

(a) Use the DJIA approach to calculate a price index for the three stocks for each period.
(b) Calculate a price index modeled after the S&P 500.
(c) Calculate the arithmetic and geometric average quarterly returns for each series of returns.

10.2. The following are the yearly returns on a portfolio for the years 1995 to 1998,

1995	1996	1997	1998
22.55%	−13.66%	−18.23%	9.34%

and the dividend and portfolio values (in millions of dollars) in 1999.

	12/31/98	3/31/99	6/30/99	9/30/99	12/31/99
Dividends	0	4	4	4	2
Value	200	195	200	200	210

(a) Calculate the rate of return on the portfolio for each quarter in 1999.
(b) If the annual return for the portfolio for 1999 is given by $(210-200)/200 = 0.05 = 5\%$, then calculate the arithmetic average return and the geometric average return for the five year interval. Explain why these are different.

10.3. The list of stocks that make up the DJIA changed on April 8, 2004. The stocks that were removed from the list along with their closing prices on April 7, 2004 are: AT&T ($19.52), International Paper ($42.40), and Eastman Kodak ($25.49). The stocks that were added to the list along with their closing prices on April 7, 2004 are: AIG ($76.25), Pfizer ($35.67), and Verizon ($37.31). The DJIA closed at 10,480.15 on April 7, 2004, and the divisor was 0.13500289. What was the new divisor on April 8, 2004?

10.4. On September 18, 1995, International Paper, a DJIA component company, had a 2-for-1 stock split. The DJIA closed at 4797.57 on September 17, 1995, and International Paper closed at $85.625. The divisor on that date was 0.3549192. What was the divisor on September 18, 1995?

10.5. Compute the Value Line Geometric Index on day 5 for the stock prices on p. 170.

10.6. What happens to the Value Line Geometric Index if all the stocks increase in price?

10.7. Verify that the Value Line Geometric Index is 100 if all the stock prices are at their original prices.

10.8. What happens to the Value Line Geometric Index if all the stocks double in price?

Running

10.9. If $x_i > 0$ for $i = 1, 2, \ldots, n$ then for $n \geq 2$, their arithmetic mean is $\bar{x} = (1/n) \sum_{i=1}^{n} x_i$.

(a) For what value of d is

$$\frac{1}{d} \left(x_1 + \cdots + x_{j-1} + \frac{1}{2} x_j + x_{j+1} + \cdots + x_n \right) = \bar{x}?$$

(b) Let a, b, and d_1 be positive, and let d_2 be such that

$$\frac{a+b}{d_1} = \frac{a + \frac{1}{2}b}{d_2}.$$

    (i) Write d_2 in terms of a, b, and d_1.
    (ii) Show that $\frac{1}{2} d_1 < d_2 < d_1$.

(c) How do the results from parts (a) and (b) relate to the computation of the DJIA?

10.10. Let S_n, T_n, U_n, and V_n be the prices of stocks S, T, U, and V on day n for the stocks on p. 170. What relationship must they satisfy so that the Value Line Geometric Index on day n is 200? 50?

10.11. From (10.1) on p. 170, show that, for any day m, where $m < t$,

$$VLG(t) = VLG(m) \left(\frac{S_1(t)}{S_1(m)} \frac{S_2(t)}{S_2(m)} \cdots \frac{S_n(t)}{S_n(m)} \right)^{1/n}.$$

10.12. Compute the percent increase in the DJIA, S&P 500, and NASDAQ Composite Index from 1980 through 2005.

10.13. Give an example of prices, $S(0), S(1), \ldots$, dividends, $D_1, D_2, \ldots$, and required rate of return, k, for a stock that satisfies (10.4) on p. 176 but not (10.3).

10.14. Verify the calculation in Example 10.7 on p. 179.

10.15. Verify the calculation in Example 10.8 on p. 181.

10.16. Verify that $\beta = 2.21$ for the Amazon.com and S&P 500 data in Example 10.9 on p. 183.

Questions for Review

- What is the DJIA? How is it computed?
- How does a stock split affect the DJIA?
- What is the Standard & Poor's 500? How is it computed?
- What is the NASDAQ Composite Index? How is it computed?
- What is the Value Line Geometric Average? How is it computed?
- What is the Value Line Arithmetic Average? How is it computed?
- What is the Value Line Theorem?
- What is the Average Quarterly Return Theorem?
- What is the Common Stock Theorem?
- What is the Constant Growth Theorem?
- What are some of the measures of the risk of a stock? How are they computed?
- What is the difference between systematic and unsystematic risk?
- What is beta? How is the beta of a stock computed? How is the beta of a portfolio computed?
- What is the CAPM?

11

Options

American options are contracts that give buyers the right, but not the obligation, to buy or sell a specified product at a specified price on or before a specified date. Options have been written for numerous products such as gold, wheat, tulip bulbs, foreign exchange, movie scripts, and stocks.[1] For example, an owner of gold might sell an option that gives the buyer the right to purchase the gold anytime in the next 30 days at a specified price. The buyer does not have to exercise that right; but if the buyer does, then the seller must sell the gold. An interesting recent example involving airline tickets can be found in [12].

Amanda Kendrick goes to a store where MP3 players are on sale for $100. However, they are sold out, and all the rain checks have been taken. Someone offers to sell her a rain check for $10, which she buys. The rain check expires in one month. In the language of options, Amanda has bought an *option* (the rain check) that has an *expiration date* of a month from now and an *exercise price* of $100 at a *premium* of $10.

Consider the following possibilities.

- The store restocks the MP3 players pricing them at $90. The option is *out of the money* because the exercise price is higher than the current price. Amanda need not exercise her option because she could buy the MP3 player from the store for $90.
- The store restocks the MP3 players pricing them at $150. The option is *in the money* because the exercise price is lower than the current price. Amanda could exercise her option by buying a $150 item for $100.
- Amanda is given an MP3 player, so she offers her option for sale. If the option is out of the money on the expiration date, then it has no value. If it is in the money, then the premium is positive and increases as the price of the MP3 player increases, but can be no larger than the difference between the price of the MP3 player and the exercise price on the expiration date.

---

[1] An advanced treatment of options and other securities is given in [14]. An elementary and entertaining discussion of options is given in [24].

The purchase price of an option is the **premium**. An option is **in the money** if exercising the option yields a profit, excluding the premium. It is **out of the money** if exercising the option is unprofitable. If the price of the asset is equal to the exercise price, then the option is **at the money**.

In this chapter we deal with stock options, which give the holder the right to buy or sell a stock for a specified price during a specified time period. Thus, the buyer of an option has the right, but not the obligation, to buy or sell the stock. The seller (writer) of an option must sell or buy the stock once the option is exercised. For an investor who holds a stock and wants to be protected against large price drops, options can guarantee a minimum price. For an aggressive investor, options can provide a great deal of leverage, that is, a small change in the price of a stock can be accompanied by a large change in the value of the options. There are option strategies that can yield a large profit for large moves in the price of a stock in either direction.

11.1 Put and Call Options

An American option gives its holder the right to buy or sell an asset for a specified price, called the EXERCISE or STRIKE PRICE, on or before a specified date, called the EXPIRATION DATE. A European option gives the same right, except the option may only be exercised on the expiration date.[2] The terms American and European refer to the type of option, not the geographical region where the options are bought or sold. A CALL OPTION gives the right to buy the asset at a specified price. A PUT OPTION gives the right to sell the asset at a specified price.

Table 11.1 shows some options information from *The Wall Street Journal* of August 22, 2000. The first column gives the name and current price of the underlying stock. We see that IBM traded at $121.44 at the end of the previous trading day. In the second column the exercise or strike prices are listed in increasing order. For IBM the exercise prices are $110, $115, $120, $125, $130, and $135. The third column lists the expiration months. For options that are listed on an exchange, the last day to trade or to exercise the options is the third Friday of the expiration month, unless the market is closed on that day. The fourth column gives the number of trades during the previous trading day for calls at that exercise price and expiration month. The fifth column gives the last transaction price for that call option. The final two columns give the volumes and last prices for the put options.

---

[2] There are several other types of options. For example, a Bermudian option can be exercised on specific dates before expiration. An Asian option calculates the exercise price as the average price of the asset over the life of the option.

Table 11.1. Option Information in August

Options	Strike	Exp.	— Call —		— Put —	
			Vol.	Last	Vol.	Last
IBM	110	Oct	14	14.88	673	2.56
121.44	115	Sep	174	8	636	1.56
121.44	115	Oct	18	11	595	4.13
121.44	115	Jan	11	16.50	546	7.13
121.44	120	Sep	570	5.13	865	3.13
121.44	125	Sep	1991	2.75	122	5.75
121.44	125	Oct	484	6	...	...
121.44	130	Sep	1136	1.13	...	...
121.44	130	Oct	984	4.13	30	11.75
121.44	135	Sep	688	0.38	5	13.50
121.44	135	Oct	608	2.69	...	...

The first line shows that the holder of that call option can buy a share of IBM at the exercise price of $110 anytime until the third Friday of October. This is called an IBM October 110 call option, which sold at $14.88 a share. Alternatively, the holder of that put option can sell a share of IBM for $110 in the same time frame. This is called an IBM October 110 put option, which sold at $2.56 a share.

Option transactions take place in terms of contracts for 100 shares. For example, if the holder of an IBM September 120 call option exercises the option, then the holder buys 100 shares of IBM at $120 a share, for a total of $12,000 (excluding commissions). Corresponding to this trade is another trade involving a writer (seller) of an IBM September 120 call option. The last price of an IBM September 120 call option is $5.13.

Notice that the last price for an IBM October 110 call option is greater than that of an IBM October 115 call option and that the last price of an October 110 put option is less than that of an October 115 put option. Also, the last price for an IBM September 115 call option is less than that of an IBM October 115 call option. We explain why these properties hold as follows.

- The holder of an IBM October 110 call option has the right to buy the stock at $110 a share, whereas the holder of an October 115 call option would pay $115 a share. This makes the IBM October 110 call option more valuable than an October 115 call option, and thus it has a higher price.
- The IBM September 115 and October 115 call options both give the right to purchase IBM stock at $115 a share, but because the IBM October 115 call option expires later than the IBM September 115 call option, there is greater opportunity for the price of IBM stock to reach a level higher than $115 during the life of the October option. Thus, the October option is more valuable, and it should have a higher price.

If the option is out of the money, then it costs less than if it were in the money. For a call option, if the stock price is less than the exercise price, then the option is out of the money. If the stock price is much less than the exercise price, then there is a very good chance that the option will expire worthless, especially if the expiration date is near. In this case, the price of the option will be low. If we look at the IBM September 135 call option, then we see that the last price of the option was 0.38, which makes sense, since the option has about one month left before expiration and is "deep" out of the money.

If the option is in the money, then the value of the option will be high. In fact, if the option is deep in the money, then a change of $1 in the price of the stock is typically matched by a change of approximately $1 in the price of the option. Note also that the closer to the expiration date of the call option, the lower the price of the option.

We quantify this by deriving a formula that gives the prices of call and put options at expiration as a function of the price of the stock by proving the following theorem.

Theorem 11.1. *The Prices of Call and Put Options Theorem.*[3]
Suppose that T is the time of expiration, that X is the exercise price, and that, at time T, the quantity $S(T)$ is the price of the stock, while $C(T)$ and $P(T)$ are the prices per share of the call and put options, respectively. Then

$$C(T) = \begin{cases} 0 & \text{if } S(T) \leq X, \\ S(T) - X & \text{if } S(T) > X, \end{cases} \tag{11.1}$$

and

$$P(T) = \begin{cases} 0 & \text{if } S(T) \geq X, \\ X - S(T) & \text{if } S(T) < X. \end{cases} \tag{11.2}$$

Proof. We prove (11.1), which is illustrated in Fig. 11.1, and leave the proof of (11.2) to you (see Problem 11.1).

Because an investor would not buy a call option if the option were out of the money, that is, $S(T) \leq X$, we have $C(T) = 0$ if $S(T) \leq X$, which is the first part of (11.1).

Now, assume that just before expiration the price of the call option is equal to $C(T)$, and the price of the underlying stock is equal to $S(T)$.

There are only three possibilities: $C(T) < S(T) - X$, $C(T) > S(T) - X$, or $C(T) = S(T) - X$. We now show that the first two are not possible, which leaves the third, which is the second part of (11.1).

If $C(T) < S(T) - X$, then an investor could buy the call option, exercise it for X, obtaining the stock, and then sell the stock for $S(T)$, making a profit of $S(T) - C(T) - X$. Such a situation could not exist for any length of time.

---

[3] This theorem makes several assumptions. For example, it assumes that there are no transaction costs, and that the call option, the put option, and the stock can be purchased at the given prices.

If $C(T) > S(T) - X$, then $C(T) + X > S(T)$, so the cost of buying the call and exercising it, $C(T) + X$, is greater than the price of the stock. Thus, an investor who wanted the stock would never buy the call option.

Thus, $C(T) = S(T) - X$ if $S(T) > X$. □

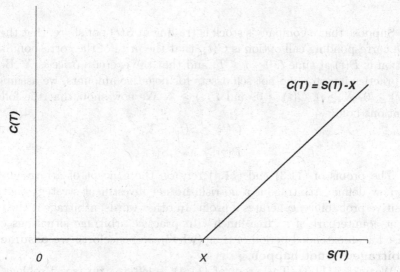

Fig. 11.1. Call-stock price relationship at expiration

There are several factors that affect the premium including the price of the underlying stock, the time to expiration, and the volatility of the returns on the stock.[4] The time to expiration is especially important. If the option is in the money, then the INTRINSIC VALUE of the option is defined as the difference between the price of the stock and the exercise price. If the option is out of the money, then the intrinsic value is zero. The time value is the difference between the premium and the intrinsic value. Thus, the premium is equal to the sum of the option's intrinsic value and time value.

For example, if the price of the stock is \$35 per share, if the exercise price for a corresponding call option is \$30, and if the call premium is \$7, then the intrinsic value is \$35 − \$30 = \$5, and the time value is \$7 − \$5 = \$2.

Once an investor has established an option position, there are three possible alternatives.

1. Liquidate the position. The writer (seller) of an option may repurchase the option in the marketplace. The buyer of an option may sell the contract. In both cases, these transactions are offsetting transactions, and the

---

[4] One measure of the volatility of a stock is the variance of the continuously compounded rate of return of the stock.

investor's option position is terminated. This is the only way in which the holder of a European option may dispose of this asset prior to expiration.

2. Hold the position. If the option expires out of the money, then the writer of the option will realize the premium as a gain (less any commissions).
3. Exercise the option. For example, the buyer of a call option buys the stock at the exercise price. The writer of the option sells the stock at the exercise price.

Suppose that a company's stock is trading at $S(t)$ per share, that the price of a corresponding call option is $C(t)$, that the price of the corresponding put option is $P(t)$ at time t, $0 \leq t \leq T$, and that the exercise price is X. Because in practice investors do not sell assets for negative amounts, we assume that $S(t) \geq 0$, $X \geq 0$, $C(t) \geq 0$, and $P(t) \geq 0$. We now show that the following relations hold:

$$C(t) \geq S(t) - X, \tag{11.3}$$

$$P(t) \geq X - S(t). \tag{11.4}$$

The proofs of (11.3) and (11.4) rely on the concept of arbitrage, which we now define. Arbitrage is a no risk, no net investment strategy that, with positive probability, generates a profit. In other words, arbitrage is the investment counterpart of a "free lunch". In practice arbitrage situations cannot exist for an extended period of time (why?), so henceforth **we assume that arbitrage cannot happen**.

We prove (11.3). The proof of (11.4) is left to you (see Problem 11.2). Clearly, if $S(t) \leq X$, then (11.3) holds because in this case $S(t) - X \leq 0$, and we must have $C(t) \geq 0$. Now, we consider the remaining case, namely $S(t) > X$. We give a proof by contradiction, that is, we assume that $C(t) < S(t) - X$ and show that this leads to an arbitrage situation, which cannot occur. If $S(t) > X$ and $C(t) < S(t) - X$, then an investor could purchase the call option for $C(t)$, exercise it for X, obtaining the stock, and then sell the stock for $S(t)$. The investor's profit is then $S(t) - C(t) - X > 0$ because $C(t) + X < S(t)$. Assuming that the transactions can be made (essentially simultaneously) and that there are no transaction costs, the investor has a risk-free, no net cost opportunity, that is, arbitrage, which cannot occur.

11.2 Adjusting for Stock Splits and Dividends

In general, option contracts are adjusted for stock splits or dividends paid of more than 10% and are not adjusted for dividends paid of less than 10%. We describe general rules for these adjustments. However, an adjustment committee, as described in the By-Laws and Rules[5] of the Options Clearing Corporation, is ultimately responsible for deciding whether an adjustment is appropriate.

---

[5] These can be found at http://www.optionsclearing.com, accessed July 1, 2006.

We first consider even stock splits. Suppose that the stock splits 2 for 1. The owner of an August 50 call option contract now owns two August 25 call option contracts. Each contract is for 100 shares, as before. In the case of even splits, the exercise price is divided by the split factor (2 in this case), and the number of contracts is multiplied by the split factor. In the case of an odd split, say 5 for 2, the situation is different. Suppose that the stock splits 5 for 2. Then the owner of an August 50 stock call now owns one August 20 stock call, but the contract is for 250 shares. In the case of an odd split, the number of contracts does not change, but the number of shares per contract is multiplied by the split factor (5/2 in this example), and the exercise price is divided by the split factor.

Adjustments are also made for stock dividends of more than 10%. The adjustments are made on the ex-dividend date.[6] Suppose that a stock declares a 25% stock dividend. Then the holder of an August 50 call option contract now owns one August 40 call option contract for 125 shares. In the case of stock dividends of more than 10%, the exercise price is divided by $1 + d$, where d is the percent of the dividend (expressed as a decimal). The number of shares per contract is multiplied by $1 + d$.

Notice that in all cases, the **total aggregate price** (the number of contracts times the exercise price per contract times the number of shares per contract) does not change. For example, in the previous example the total aggregate price is $1 \times 40 \times 125 = \$5,000$, the same as the original price of $1 \times 50 \times 100 = \$5,000$. Table 11.2 summarizes the different possibilities.

Table 11.2. Effects of Stock Splits and Dividends on Option Prices

	Strike Price	Number of Contracts	Number of Shares/Contract
Even split (2 for 1)	25	2	100
Odd split (5 for 2)	20	1	250
Dividend > 10%	40	1	125
Dividend ≤ 10%	50	1	100 (no change)

---

[6] When a company declares a dividend, it sets a date by which a stock buyer must be on the company's books as a shareholder in order to receive the dividend. After this date is set, the stock exchanges set the ex-dividend date, which is usually two business days before the record date. If a buyer purchases a stock on its ex dividend date or later, the buyer does not receive the next dividend payment; the seller does. If a buyer purchases the stock before the ex-dividend date, the buyer receives the dividend.

11.3 Option Strategies

We now consider some common option strategies. Of particular emphasis are the regions of profit and loss, the break-even points, and the maximum profit and loss. In the following, we assume that there are no transaction costs and that the specified prices can be realized.

11.3.1 Buying Calls

Suppose that the price of the stock at time t is $S(t)$, the price of the call is $C(t)$, and the exercise price is X, where $0 \leq t < T$. If $X > S(t)$, then the option is out of the money, and the option should not be exercised, so we consider the case $X \leq S(t)$. From 11.3, we have $C(t) \geq S(t) - X$, so the holder of the call option makes $C(t)$ by selling the option, thereby liquidating the position. If the holder of the option exercises it and then immediately sells the stock, then the holder earns $S(t) - X \leq C(t)$. Therefore, it is never in the interest of the holder to exercise a call option prior to expiration.

We now look at what happens at expiration. Suppose that Tom Kendrick is bullish[7] on the stock of HIGH Corp. The stock is currently at \$48 a share, and the September 50 call option is at \$2 a share. Because $X = 50$, the option is out of the money. Tom buys one September 50 call option, so his cost, C, is \$2 per share. If he sells the contract at or before expiration, then he makes a profit if and only if the current option price is greater than \$2. If he waits until expiration, then the price of the contract per share, $C(T)$, is given by (11.1), which in this case is

$$C(T) = \begin{cases} 0 & \text{if } S(T) \leq 50, \\ S(T) - 50 & \text{if } S(T) > 50. \end{cases}$$

Tom's profit or loss per share is $C(T) - C$. He makes a profit if $C(T) > C$, that is, if $S(T) - 50 > 2$ or $S(T) > \$52$. He makes a loss if $C(T) < C$, that is, if $S(T) < \$52$. His total profit is $100(C(T) - C)$, which, if negative, represents a loss. His potential profit is unlimited because at expiration each \$1 increase in the price of the stock is matched by a corresponding \$1 increase in the price of the option. His potential loss is limited to his initial investment of \$200 (\$2 × 100 shares). Thus, we have Fig. 11.2, which shows his total profit or loss on 100 shares.

From (11.1) we have

$$C(T) = \begin{cases} 0 & \text{if } S(T) \leq X, \\ S(T) - X & \text{if } S(T) > X. \end{cases}$$

---

[7] Someone is bullish on a stock if that person believes that the stock will increase in price. Someone is bearish if that person believes that the stock will decrease in price.

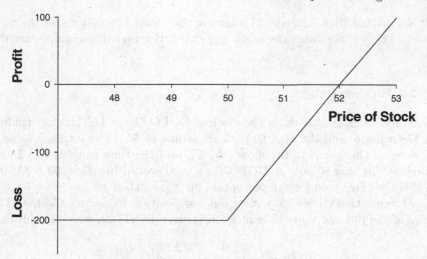

Fig. 11.2. Tom Kendrick's total profit at expiration

If C is the price paid for the call option per share, then the profit per share, $C(T) - C$, is

$$\text{Profit per share} = \begin{cases} -C & \text{if } S(T) \leq X, \\ S(T) - X - C & \text{if } S(T) > X, \end{cases}$$

so the profit-loss diagram for buying calls is of the form shown in Fig. 11.3.

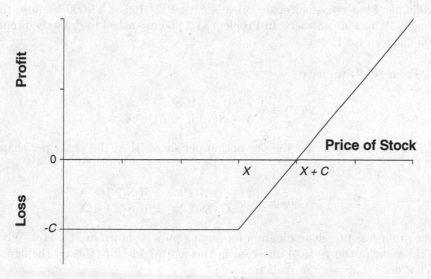

Fig. 11.3. Profit per share at expiration when buying calls

Notice that Figs. 11.2 and 11.3 are not the same. The first shows the total profit. The second shows the profit per share. However, their shapes are the same.

11.3.2 Buying Puts

Helen Kendrick is bearish on the stock of GOLO Corp. The stock is trading at $78 a share, and the May 80 puts are selling at $5. (The intrinsic value is $2 because the put is in the money by $2, and the time value is $3.) Helen purchases 10 May 80 puts on GOLO, for a total cost of $10 \times \$5 \times 100 = \$5,000$. If P is the price paid for the put option per share, then we have $P = 5$.

At expiration, Helen may close out her position by selling the put. The price of the put per share at expiration is given by (11.2), which in this case is

$$P(T) = \begin{cases} 0 & \text{if } S(T) \geq 80, \\ 80 - S(T) & \text{if } S(T) < 80. \end{cases}$$

Helen's profit or loss per share is $P(T) - P$. She makes a profit if $P(T) > P$, that is, if $80 - S(T) > 5$ or $S(T) < \$75$. She makes a loss if $P(T) < P$, that is, if $S(T) > \$75$. Her total profit is $10 \times 100(P(T) - P)$, which, if negative, represents a loss. Her maximum profit occurs at a stock price of zero. (This is actually possible as one of the authors discovered.) At this point the put can be sold for $80, and Helen's total profit is $10 \times \$80 \times 100 - \$5,000 = \$75,000$. As with Tom, Helen's maximum loss is her original investment, in this case $5,000. Notice that Helen paid only $5,000 to establish her position, but her profit per $1 decrease in share price is $\$1 \times 10 \times 100 = \$1,000$ because she bought 10 option contracts. In Problem 11.3 you are asked to draw the profit-loss diagram for Helen.

From (11.2) we have

$$P(T) = \begin{cases} 0 & \text{if } S(T) \geq X, \\ X - S(T) & \text{if } S(T) < X. \end{cases}$$

If P is the price paid for the put option per share, then the profit per share, $P(T) - P$, is

$$\text{Profit per share} = \begin{cases} -P & \text{if } S(T) \geq X, \\ X - S(T) - P & \text{if } S(T) < X. \end{cases}$$

The profit-loss per share diagram for buying puts is shown in Fig. 11.4. What is the value of the vertical intercept in this figure? Identify this on the figure. What does this mean in realistic terms?

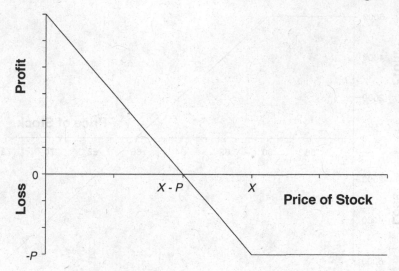

Fig. 11.4. Profit per share at expiration when buying puts

11.3.3 Writing Calls

A person who is bearish on the stock of a company may write calls on the stock. The writer of the call is obliged to sell the stock at the exercise price if the call is exercised. There are two different situations, depending on whether the writer of the call owns the stock or not. We consider the more risky situation first.

Writing Uncovered Calls

Uncovered calls are calls in which the writer of the call does not own the underlying stock. This is considered one of the most risky option strategies because the uncovered call writer has unlimited loss potential. The writer of an uncovered call is subject to a margin requirement.

The stock of DROP Corp. is currently trading at $63 a share. A September 60 call ($3 in the money) is selling for $6. Hugh Kendrick writes 10 September 60 calls on DROP. His premium for the transaction is $10 \times \$6 \times 100 = \$6,000$. He does not own the stock, so the calls are uncovered. Hugh keeps the entire premium if DROP closes at or below the exercise price of $60 at expiration. He loses $1 \times 10 \times 100 = \$1,000$ times $(S(T) - 60)$ if the price of the stock is above $60 at expiration but keeps the premium of $6,000. Thus, his break-even point is $66. If $S(T) > 66$, then he has a net loss of $\$1,000(S(T) - 66)$. Thus, Hugh has unlimited loss potential. Figure 11.5 shows Hugh's profit or loss at expiration.

If DROP increases in price, then Hugh may receive a maintenance call. He can liquidate his position at any time by buying 10 September 60 calls on DROP.

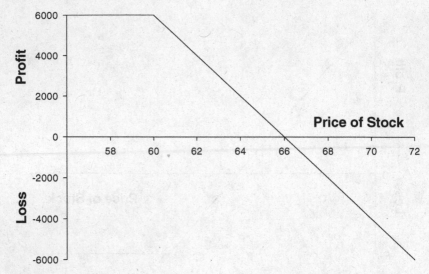

Fig. 11.5. Hugh Kendrick's position at expiration for writing uncovered calls

In general, we have

$$\text{Profit per share} = \begin{cases} C & \text{if } S(T) \leq X, \\ X - S(T) + C & \text{if } S(T) > X, \end{cases}$$

where C is the price per share received for the call, and X is the exercise price per share. Fig. 11.6 shows the profit-loss per share diagram for this case.

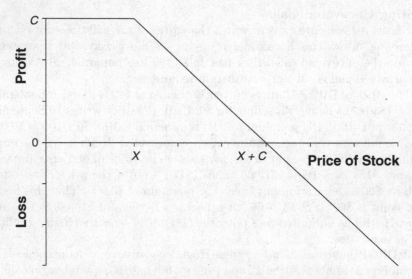

Fig. 11.6. Per share uncovered call writing position at expiration

Writing Covered Calls

Covered calls differ from uncovered calls in that the writer of covered calls owns the appropriate number of shares of the underlying stock. Referring to the last example, if Hugh had bought 1,000 shares of DROP at $62 a share prior to writing the calls, then he could supply the shares if the option were exercised.

We now consider Hugh's initial position if he wrote covered calls on DROP and had previously purchased 1,000 shares at $62 a share. He paid $62,000 for the stock and received $6,000 for the call options. Therefore, the net cost of establishing the position is $56,000. This situation is more complicated than the uncovered call case. Note that Hugh keeps the $6,000 premium if DROP closes at or below $60 at expiration. However, if it closes at $56 a share at expiration, then he loses $1,000(62 − 56) = $6,000 due to the decrease in the value of his purchased shares, making a zero net profit. If the stock closes below $56, then he has a net loss. His maximum profit occurs if DROP closes at or above $60 at expiration. At that point his profit is $6,000 for the call option and −$2,000 for the cost of the shares purchased, for a net profit of $4,000. Hugh's profit-loss diagram is shown in Fig. 11.7.

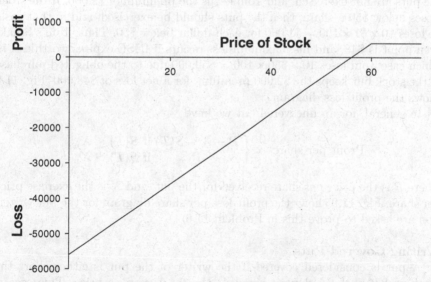

Fig. 11.7. Hugh Kendrick's covered call writing position at expiration

In general, for the writer of a covered call, we have

$$\text{Profit per share} = \begin{cases} S(T) - S(0) + C & \text{if } S(T) \le X, \\ X + C - S(0) & \text{if } S(T) > X, \end{cases}$$

where $S(0)$ is the price paid for the stock, C is the price per share received for the call, and X is the exercise price per share. You are asked to prove this in Problem 11.5 and to draw the corresponding profit-loss diagram.

11.3.4 Writing Puts

A person who is bullish on a stock may write puts on the stock. The writer of a put is obligated to buy the stock at the exercise price if the option is exercised. As with the writing of calls, there are two different cases when writing puts.

Writing Uncovered Puts

In this case, the writer of the put does not have cash on deposit equal to the exercise price times the number of shares written and is not short the stock.[8] The writer of an uncovered put is subject to a margin requirement.

Tom Kendrick is bullish on the stock of RISE Corp. Tom writes 10 September 50 puts on RISE at $2 per share. His premium for the transaction is $10 \times \$2 \times 100 = \$2,000$. Tom does not have cash on deposit equal to the total exercise price ($10 \times \$50 \times 100 = \$50,000$) and is not short the stock, so this is an uncovered put, and he is subject to a margin requirement.

If the stock of RISE closes at a price greater than $50 at expiration, then the puts are not exercised, and Tom keeps the premium of $2,000. If the stock closes below $50 a share, then the puts should be exercised, and in that case, he loses $10 \times \$1 \times 100 = \$1,000$ for each dollar below $50. Thus, Tom's break-even point is $48, and his maximum loss occurs if RISE expires worthless, in which case Tom loses $10 \times \$50 \times 100 = \$50,000$ due to the obligated purchase of the stock but keeps the $2,000 premium, for a net loss of $48,000. Fig. 11.8 shows the profit-loss diagram.

In general, for an uncovered put we have

$$\text{Profit per share } = \begin{cases} P - X + S(T) & \text{if } S(T) \le X, \\ P & \text{if } S(T) > X, \end{cases}$$

where P is the price per share received for the put, and X is the exercise price per share. Fig. 11.9 shows the profit-loss per share diagram for the put writer. You are asked to prove this in Problem 11.6.

Writing Covered Puts

A put is considered covered if the writer of the put is either short the stock or has cash on deposit equal to the total exercise price. This case is more complicated than the case with uncovered puts, and we shall see that in this case there is no limit to the potential loss if the writer is short the stock.

Using the last example, suppose that Tom Kendrick is short 1,000 shares of RISE, having borrowed the 1,000 shares from his investment firm and selling

---

[8] The expression "short the stock" means "having a short position in the stock".

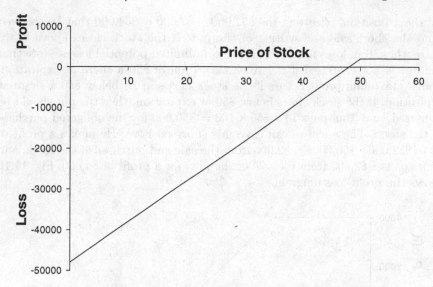

Fig. 11.8. Tom Kendrick's uncovered put writing position at expiration

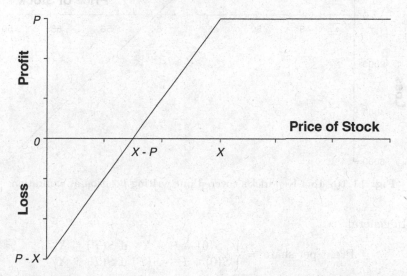

Fig. 11.9. Uncovered put per share writing position at expiration

them for $52 a share. In this case, Tom has $52,000 on deposit,[9] and thus he has a covered position. The exercise price is $50, and Tom writes 10 September 50 puts on RISE for $2 a share.

In this case, Tom loses $10 \times \$1 \times 100 = \$1,000$ for each dollar above $52 at which the stock closes because he is short the stock. Tom's break even point is $54. At this point Tom has to pay $10 \times \$54 \times 100 = \$54,000$ to cover

---

[9] This does not include any margin paid to the broker.

his short position, offsetting the $52,000 + $2,000 = $54,000 that he received from the short sale and writing of the put. If the stock closes above $54 a share, then Tom loses money, and he has unlimited potential losses. Note that the put is not exercised if the stock closes above $50 a share at expiration. Tom's maximum profit occurs if the stock closes at or below $50 a share at expiration. If the stock closes below $50 at expiration, then the put should be exercised, and Tom pays $10 \times \$50 \times 100 = \$50,000$ for the obligated purchase of the stock. These shares can cover his short position. He makes a profit of $10 \times (\$52,000 - \$50,000) = \$2,000$ from the sale and purchase of the stock, and he keeps the $2,000 from the sale of the put, for a profit of $4,000. Fig. 11.10 shows the profit-loss diagram.

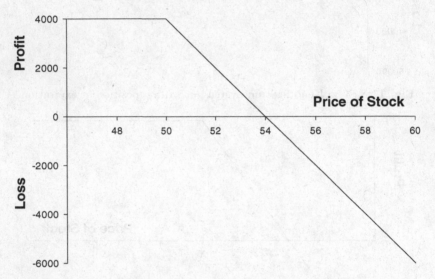

Fig. 11.10. Tom Kendrick's covered put writing position at expiration

In general,

$$\text{Profit per share} = \begin{cases} S(0) + P - X & \text{if } S(T) \leq X, \\ S(0) + P - S(T) & \text{if } S(T) > X, \end{cases}$$

where $S(0)$ is the price per share received for the short sale, P is the price per share received for the put, and X is the exercise price per share, and we assume that $S(0) + P > X$. The profit-loss per share diagram for the writing of covered puts, where the writer is short the stock, is shown in Fig. 11.11. You are asked to prove this in Problem 11.7. What is the value of the vertical intercept in this figure? Identify this on the figure. What does this mean in realistic terms?

The maximum profit occurs when the price of the stock at expiration is less than or equal to X. The break-even point occurs when the stock price is $S(0) + P$, and there are unlimited potential losses.

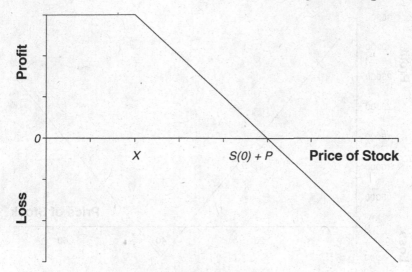

Fig. 11.11. Covered put per share writing position at expiration

11.3.5 Straddles

A straddle is the purchase of a call and a put or the sale of a call and a put on a stock, with the exercise price and expiration date being the same for both options. The buyer of a straddle expects a large change in the price of the stock but is unsure of the direction of the change. If there is a large increase in the price of the stock, then the buyer of the straddle may profit from the call option, whereas if there is a large decrease in the price of the stock, then the buyer may profit from the put option. Conversely, the writer of a straddle expects little or no change in the price of the stock and may profit from the premiums received from the sale of the options. We study the case where the investor buys a straddle. The case of the writer of a straddle is left as an exercise (see Problem 11.8).

Suppose that Helen Kendrick believes that the stock of MOVE Corp. is due for a large change in price but is unsure as to the direction of the change. MOVE is currently selling at \$40 a share. The June 40 call options on MOVE are selling at \$3 a share, and the June 40 put options are selling at \$2 a share. Helen buys 10 June 40 calls and 10 June 40 puts on MOVE, paying a total of $(10 \times \$3 \times 100) + (10 \times \$2 \times 100) = \$5,000$. If MOVE is below \$40 a share at expiration, then Helen profits from the put position. If MOVE is above \$40 a share at expiration, then she profits from the call position. The profit-loss diagram for Helen's position is shown in Fig. 11.12.

Helen breaks even if MOVE closes at \$45 or at \$35 at expiration. She loses if the price is between \$35 and \$45 at expiration. Her maximum loss occurs if MOVE closes at the exercise price of \$40 a share at expiration, in which case she loses the entire premium of \$5,000. Her potential profit from the call

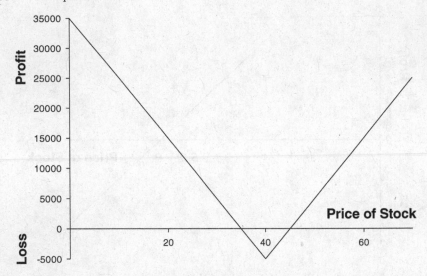

Fig. 11.12. Helen Kendrick's straddle position at expiration

option is unlimited, whereas the maximum profit from the put option occurs if the stock expires worthless, in which case the put option yields a profit of $40,000, and her total profit is $40,000 − $5,000 = $35,000.

In general,

$$\text{Profit per share } = \begin{cases} X - S(T) - C - P \text{ if } S(T) \leq X, \\ S(T) - X - C - P \text{ if } S(T) > X, \end{cases}$$

where C is the price paid per share for the call option, P is the price paid per share for the put option, and X is the exercise price per share. The profit-loss per share diagram for the buyer of a straddle is shown in Fig. 11.13. You are asked to prove this in Problem 11.9. What is the value of the vertical intercept in this figure? What is the minimum value? Identify these on the figure. What do these mean in realistic terms? Does the function have a maximum value?

There are many other option strategies. The ones that we have discussed are very common.

11.4 Put-Call Parity Theorem

Consider a call option with price C, and a put option with price P, where both are European options written on the same stock. (A European option may not be exercised prior to expiration.) Both options have the same exercise price X and expiration date T. At time $t = 0$ we form two portfolios. The first consists of the call option and a risk-free zero-coupon bond, with a face value of X, maturing at time T. The second portfolio consists of the stock and the put option.

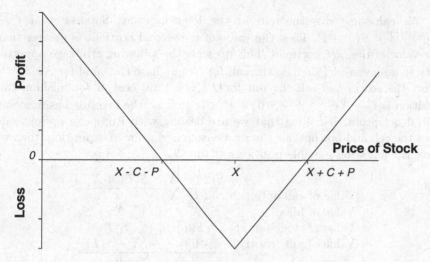

Fig. 11.13. Straddle position at expiration

The following table compares the values of the two portfolios at time T, where $S(T)$ is the price of the stock at time T.

		$S(T) \leq X$	$S(T) > X$
Portfolio 1:	Value of call option	0	$S(T) - X$
	Value of bond	X	X
	Total value	X	$S(T)$
Portfolio 2:	Value of stock	$S(T)$	$S(T)$
	Value of put option	$X - S(T)$	0
	Total value	X	$S(T)$

It follows that, regardless of the stock price at time T, the values of the two portfolios are the same at time T. Therefore, by our no arbitrage assumption, the initial costs of creating the two portfolios should be the same.

The cost of creating the first portfolio is $C + Xe^{-i^{(\infty)}T}$, where $i^{(\infty)}$ is the continuously compounded risk-free rate of return. The cost of creating the second portfolio is $S(0) + P$, where $S(0)$ is the price of the stock at time 0. This yields the following theorem.

Theorem 11.2. *The Put-Call Parity Theorem.*
Suppose that C is the price of a call option, that P is the price of a put option (where both are European options written on the same stock), that both options have the same exercise price X and expiration date T, that $S(0)$ is the price of the stock at time 0, and that the stock pays no dividends. Then

$$C + Xe^{-i^{(\infty)}T} = S(0) + P,$$

where $i^{(\infty)}$ is the continuously compounded risk-free rate of return.

We can view this theorem in the following way. Suppose that $C + Xe^{-i(\infty)T} < S(0) + P$. Then the value of the second portfolio is greater than the value of the first portfolio. This presents the following arbitrage opportunity to an investor: Purchase the call for C, purchase the bond for $Xe^{-i(\infty)T}$, short the stock, and sell the put for P. The total cost of establishing this position is $C + Xe^{-i(\infty)T} - S(0) - P < 0$, that is, the investor has made an initial net profit. Recalling that we are dealing with European options and that the call and put options cannot be exercised prior to expiration, then we have the following possible positions at time T.

	$S(T) \geq X$	$S(T) < X$
Value of call option	$S(T) - X$	0
Value of bond	X	X
Value of stock (short)	$-S(T)$	$-S(T)$
Value of put (short)	0	$-(X - S(T))$
	0	0

Thus, the value of the portfolio at time T is 0, regardless of the price of the stock, $S(T)$. Since there is an initial net profit in setting up the portfolio, there is a risk-free, guaranteed profit. We assume that this arbitrage situation cannot exist.

Similarly, if $C + Xe^{-i(\infty)T} > S(0) + P$, then an investor can short the call for C, short the bond for $Xe^{-i(\infty)T}$, purchase the stock for $S(0)$ and purchase the put for P. In Problem 11.10 you are asked to verify that this leads to an arbitrage opportunity.

The following example illustrates the use of the Put-Call Parity Theorem.

Example 11.1. HOT1 is currently selling at \$95 per share, pays no dividends, and has call and put options expiring at time $T = 0.5$ (six months) with the same exercise price of \$100. The call costs \$5 and the put costs \$8, and the continuously compounded risk-free rate of return is 10%.[10] Calculate the costs of creating Portfolios 1 and 2 discussed on p. 209.

Solution.
Portfolio 1.

$$C + Xe^{-i(\infty)T} = 5 + 100e^{-0.10 \times 0.5} = 5 + 95.12 = 100.12.$$

Portfolio 2.
$$S(0) + P = 95 + 8 = 103.$$

Thus, the cost of creating the first portfolio is less than the cost of creating the second portfolio. We take advantage of this arbitrage situation by "buying" the first portfolio and "selling" the second portfolio. That is, we purchase the

---

[10] Note that the put is in the money and thus has a higher premium than the call.

call for \$5, we purchase the bond for \$95.12, we short the stock for \$95, and we sell the put for \$8. This represents a net profit of \$2.88. (Note that when we short a stock, we must pay any dividends to the lender of the stock. This is one of the reasons that we assumed that the stock pays no dividends.) At time T (six months) we have the following values for the portfolios.

	$S(T) \leq X$	$S(T) > X$
Value of call option	0	$S(T) - 100$
Value of bond	100	100
Value of stock (short)	$-S(T)$	$-S(T)$
Value of put (short)	$-100 + S(T)$	0
Total	0	0

We have achieved the desired result. We make a net profit of \$2.88 in creating the position, and the value of the portfolio at time T is 0, independent of the price of the stock, $S(T)$. $\triangle$

Let us review some of the assumptions made. We assumed that we could purchase a risk-free bond with a face value of X that matures at time T. Since the bond is risk-free, this excludes such investments as corporate bonds, municipal bonds, and callable bonds. We also assumed that the portfolio could be created at the stated values. For example, the stock, the bond, and the call and put options could be purchased (or sold short) at the given prices, in effect simultaneously. This is a liquidity assumption. We assumed that the options were European, so they cannot be exercised prior to expiration. We previously pointed out that, at least in principle, American call options are never exercised prior to expiration, but there are cases where it makes sense to exercise put options. The following (extreme) case illustrates this point: Suppose that a put is bought on a company that files for bankruptcy, after which the stock price is zero. Then it is obvious that the put should be exercised, since the price of the stock can go no lower. A similar argument holds if the price of the stock falls to a low non-zero price.

11.5 Hedging with Options

Hedging involves offsetting the risk of a given investment or portfolio by investing in a different asset. For example, an investment in a particular stock may be accompanied by the purchase of a put option.

Example 11.2. Wendy Kendrick purchases 100 shares of DROP at \$30 a share. She is worried that the stock may decrease in price and wants to limit her potential loss. She purchases one put option on DROP with an exercise price of \$25. What happens if the price of DROP falls below \$25?

Solution. If the stock price falls below \$25, then Wendy can exercise the put option and limit her loss due to the decrease in stock price to $100 \times (\$30 -$

$25) = \$500$. This is an example of a PROTECTIVE PUT. Note that she pays $30 a share for the stock, pays for the put option, and pays commissions for each transaction. $\triangle$

Now, consider the following example: At the beginning of the year, GOUP trades at $50 a share and it can move to either $25 or $100 per share by the end of the year. (Assume that these are the only possibilities.) Suppose that the continuously compounded risk-free rate of return is 5% and that a call option for GOUP has an exercise price of $75. At expiration the call price is $25 if the price of the stock is $100 and $0 if the price of the stock is $25. This is illustrated in the following table.

	$S(T) \leq X$	$S(T) > X$
$S(T)$	25	100
$C(T)$	0	25

We illustrate hedging in the following way. We wish to create a portfolio by purchasing Δ shares of the stock and borrowing the present value of B dollars discounted at 5% compounded continuously. This portfolio is to have the same value as another portfolio consisting of one call option. In order to owe B at the end of the year, we must borrow $Be^{-i^{(\infty)}}$. In order to create the desired portfolio, the following equation must be satisfied

$$\Delta S(T) - B = C(T),$$

where $S(T)$ is the price of the stock at the end of the year and $C(T)$ is the price of the call on the expiration date. This leads to the system of equations

$$25\Delta - B = 0, \text{ and}$$
$$100\Delta - B = 25.$$

Solving for Δ and B gives $\Delta = 1/3$ and $B = 25/3$. Thus, we purchase 1/3 share of GOUP at $50 a share and borrow $(25/3)e^{-0.05} = \$7.93$. (Note that we have assumed that we can buy fractional shares of the stock.) At the end of the year we have the following values for the portfolios.

	$S(T) \leq X$	$S(T) > X$
Portfolio 1: One call option	0	25
Total	0	25
Portfolio 2: 1/3 share of GOUP	8.33	33.33
Loan Repay	−8.33	−8.33
Total	0	25

Thus, we have replicated the first portfolio with 1/3 share of GOUP and a loan of $7.93 at 5% compounded continuously. What does this tell us? Using the usual arbitrage arguments, the costs of creating the two portfolios must be the same. The cost of $1/3$ share of GOUP is $1/3 \times \$50 = \$16.67$. The cost of the loan is $-\$7.93$. Thus, the total cost to set up this portfolio is $\$8.74$. Because Portfolio 1 consists of one call option, the price of a call option on GOUP should be $\$8.74$.

We now generalize the hedging method. We have a stock whose current price is $S(0)$, with a call expiring at time $T > 0$ with exercise price X. The continuously compounded risk-free rate of return is $i^{(\infty)}$. The stock can either increase in price to S_U or decrease in price to S_D at expiration, and we assume that these are the only two possibilities. We wish to create a portfolio consisting of Δ stock shares and cash in a risk-free investment with a future value of B. Then the value of our hedged portfolio agrees with the value of the call option, regardless of the price of the stock at time T, if the following two equations are satisfied:

$$\Delta S_U - B = C_U,$$
$$\Delta S_D - B = C_D,$$

where S_U and S_D are the possible values of the stock at expiration, and C_U and C_D are the corresponding prices of the call options. Solving these equations for Δ and B, we find that

$$\Delta = \frac{C_U - C_D}{S_U - S_D},$$

the slope of the line passing through the points (S_D, C_D) and (S_U, C_U), and that

$$B = \frac{C_U S_D - C_D S_U}{S_U - S_D}.$$

In our example, we have $S_D = 25$, $S_U = 100$, $C_D = 0$, and $C_U = 25$.

We now consider a slightly different problem: that of hedging with a portfolio constructed by purchasing Δ shares of the stock and selling one call option. We make the same assumptions and use the same notation as before. We call the quantity

$$\frac{C_U - C_D}{S_U - S_D}$$

the HEDGE RATIO.

We have the following theorem.

Theorem 11.3. *The Hedge Ratio Theorem.*
The hedge ratio is the ratio of stock held for every one call option sold so that the value of the portfolio is unaffected by the price of the stock.

Proof. For the proof of the theorem, we simply show that the value of the portfolio is the same for the two values S_U and S_D. When the price of the stock is S_U, then the value of the portfolio is

$$\Delta S_U - C_U = \frac{C_U - C_D}{S_U - S_D} S_U - C_U$$
$$= \frac{C_U S_D - C_D S_U}{S_U - S_D}.$$

When the price of the stock is S_D, then the value of the portfolio is

$$\frac{C_U - C_D}{S_U - S_D} S_D - C_D = \frac{C_U S_D - C_D S_U}{S_U - S_D}.$$

This completes the proof of the theorem.   □

An important feature of the Hedge Ratio Theorem is that the result is independent of the probabilities that the stock increases in price from $S(0)$ to S_U or decreases in price from $S(0)$ to S_D.

We note that in the previous example we had $S_U = 100, S_D = 25, C_U = 25$, and $C_D = 0$, so for that example,

$$\Delta = \frac{25 - 0}{100 - 25} = \frac{1}{3},$$

that is, we should hold 1/3 share of stock for every call option sold.

Referring to this example, suppose that the call option is overpriced, say $C = 10$ instead of 8.74. Then we have $S(0) = 50$, $S_U = 100$, $S_D = 25$, $C_U = 25$, $C_D = 0$, and $i^{(\infty)} = 0.05$.

We can use arbitrage to make a profit as follows: write three call options at 10, purchase one share at 50, and borrow \$20 at 5% compounded continuously. Our initial cash flow is $30 - 50 + 20 = 0$, so there is no cost of financing the portfolio. Let us see what happens at expiration:

	$S(T) = 25$	$S(T) = 100$
Write three options	0.00	−75.00
Purchase one share	25.00	100.00
Borrow \$20 at 5% compounded continuously	−21.03	−21.03
	3.97	3.97

We have a riskless investment. The initial investment was zero, and we have a profit of \$3.97, independent of the price of the stock at expiration. Note that to get the desired profit, we simply used the hedge ratio of 1/3 (one-third share of stock for every call option sold) and borrowed enough money at the risk-free interest rate to have no initial cost.

11.6 Modeling Stock Market Prices

In his doctoral thesis [2] written in 1900, Bachelier laid the groundwork for the study of stock market pricing. A discussion of different approaches is given in [6]. Much of the material in this and the next section follows arguments developed in [23].

We now discuss the pricing of options. Because the price of an option depends, in large part, on the price of the underlying stock, we first discuss the pricing of stocks, which in turn depends on how stock market prices are modeled.

Suppose that we follow the price of a stock over time. Let $S(t)$, $0 \leq t < \infty$, denote the price of the stock at time t.

One model of these prices is the BROWNIAN MOTION MODEL. With this model we assume that if $0 \leq t_0 < t_1 < \cdots < t_{n-1} < t_n$, $n \geq 1$, then $S(t_{i+1}) - S(t_i)$, $i = 0, 1, \ldots, n - 1$, has a normal distribution with mean $\mu(t_{i+1} - t_i)$ and variance $\sigma^2(t_{i+1} - t_i)$ and that the random variables $S(t_1) - S(t_0), \ldots, S(t_n) - S(t_{n-1})$ are independent. We assume that $S(t_0)$ is known.

Thus, the change in the price of the stock has a normal distribution, with the mean and variance of the price change linear in the length of the interval. The price change is independent of the previous prices. One of the drawbacks of this model is that the change $S(t_{i+1}) - S(t_i)$ does not depend on the price $S(t_i)$ at time t_i, $i = 0, 1, \ldots, n - 1$. Thus, for example, the probabilities that a stock with price $S(t_n) = \$2.00$ at time t_n and that a different stock with price $S(t_n) = \$100.00$ at time t_n both increase in price by \$1.00 in the next time period are the same.

A modified version of this model considers the price ratios $S(t_{i+1})/S(t_i)$, rather than differences. This is the GEOMETRIC BROWNIAN MOTION MODEL. For this model it is assumed that the ratios $S(t_1)/S(t_0), \ldots, S(t_n)/S(t_{n-1})$ are independent and that the random variable $\ln(S(t_{i+1})/S(t_i))$ is a normal random variable with mean $\mu(t_{i+1}-t_i)$ and variance $\sigma^2(t_{i+1}-t_i)$, $i = 0, \ldots, n-1$. Note that, with this model, $S(t_{i+1})/S(t_i)$ has a lognormal distribution, and if two stocks have this distribution with the same parameter σ^2, then the likelihood that each stock increases in price by the same percent is the same for each stock.

The BINOMIAL MODEL of stock prices is a discrete model. We assume that a stock trades during a fixed time interval $[0, T]$, and we divide the interval into n equally spaced subintervals

$$\left[0, \frac{T}{n}\right], \left(\frac{T}{n}, \frac{2T}{n}\right], \ldots, \left(\frac{(n-1)T}{n}, T\right].$$

We define $t_0 = 0, t_1 = T/n, \ldots, t_n = T$. This is called the n-step model.

The model is the following: If the stock price is $S(t_i)$ at time t_i, then we assume that it may increase to price $S_U(t_{i+1})$ with probability $P(t_i)$ at time t_{i+1} or decrease to price $S_D(t_{i+1})$ with probability $1 - P(t_i)$.[11] We also assume that the price changes $S(t_1) - S(t_0)$, $S(t_2) - S(t_1), \ldots, S(t_n) - S(t_{n-1})$ are independent. Then at each time t_{i+1} the stock price may be at either $S_U(t_{i+1})$ or $S_D(t_{i+1})$, and these values depend only on the past price $S(t_i)$ and the probability $P(t_i)$. Our model (a recombining tree) contains the assumption that the value of a stock after a decrease followed by an increase is the same as that of the stock after an increase followed by a decrease.

We begin with a simple model that is used in practice. This is a STAN-DARD BINOMIAL TREE, in which it is assumed that $P(t_i)$ and the ratios $S_U(t_{i+1})/S(t_i)$ and $S_D(t_{i+1})/S(t_i)$ are constant for $i = 0, \ldots, n - 1$. These constants are denoted by p, u, and d, respectively, and u and d are called the **up and down ratios**.

Example 11.3. Confirm that Table 11.3 is the standard binomial tree for $n = 4$, $u = 1.02$, $d = 0.98$, and $S(t_0) = 100$. All values have been rounded off to the nearest 0.01.

Table 11.3. Standard Binomial Tree for Example 11.3

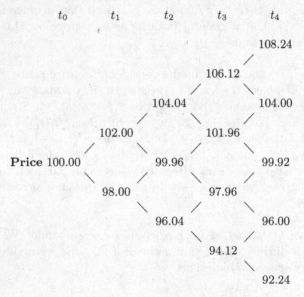

---

[11] The probabilities $P(t_i)$ and $1 - P(t_i)$ are called transition probabilities.

Solution. This tree can be constructed in the following way: The nodes in the first column are obtained from $100 \times 1.02 = 102$, and $100 \times 0.98 = 98$. The node in the third column with a price of 99.96 represents the price of the stock after two time periods with either an increase followed by a decrease or a decrease followed by an increase. Thus, $100 \times 1.02 \times 0.98 = 100 \times 0.98 \times 1.02 = 99.96$. (Note that, contrary to intuition, the price is not 100, the initial price.) At time t_2 the stock can have three possible prices: 96.04, 99.96, or 104.04, corresponding to two decreases in a row, one increase and one decrease (in either order), or two increases in a row. $\triangle$

Because the probabilities are p and $1 - p$ at each step, and we have a binomial model, $P(S(t_1) = 102.00) = p$ is the probability that there is an increase in price from time t_0 to time t_1, and $P(S(t_1) = 98.00) = 1 - p$ is the probability of a decrease. We can calculate the other probabilities in the same manner. For example, $P(S(t_2) = 96.04) = (1 - p)^2$, and $P(S(t_2) = 99.96) = p(1 - p) + (1 - p)p = 2p(1 - p)$. What is $P(S(t_2) = 104.04)$?

We have not yet specified the value of p. We first consider a one-step model. Because $S(t_1)$ is a random variable, we can compute its expected value, which is $pS_U(t_1) + (1 - p)S_D(t_1)$. Here $S_U(t_1)$ is the price following an increase, and $S_D(t_1)$ is the price following a decrease.

Now, we wish to determine the p that will yield a given (continuously compounded) rate of return. We let $i^{(\infty)}$ be the required rate of return per year. Then, because the initial price was $S(t_0)$, we want the expected value of $S(t_1)$ to be $S(t_0)e^{i^{(\infty)}\delta t}$, where $\delta t = t_1 - t_0$. Thus, we need $pS_U(t_1) + (1 - p)S_D(t_1) = S(t_0)e^{i^{(\infty)}\delta t}$.

Solving for p gives

$$p = \frac{S(t_0)e^{i^{(\infty)}\delta t} - S_D(t_1)}{S_U(t_1) - S_D(t_1)}.$$

For the standard tree, $P(t_i) = p$, $S_U(t_{i+1})/S(t_i) = u$, and $S_D(t_{i+1})/S(t_i) = d$ for all i. Thus, we have

$$p = \frac{e^{i^{(\infty)}\delta t} - d}{u - d}. \tag{11.5}$$

In this case we need $pu + (1 - p)d = e^{i^{(\infty)}\delta t}$. This leads to the following result.

Theorem 11.4. *For an n-step standard binomial tree with initial price $S(t_0)$, transition probabilities p and $1 - p$, and up and down ratios u and d, respectively, suppose that the expected value of $S(t_1)$ is $S(t_0)e^{i^{(\infty)}\delta t}$, where $\delta t = t_1 - t_0$. Then the expected value of $S(t_i)$ is $S(t_0)e^{i^{(\infty)}i\delta t}$, for $i = 0, 1, \ldots, n$. Thus, the expected value of the stock price is equal to the return on an investment of $S(t_0)$ with a continuously compounded rate of return of $i^{(\infty)}$ per year.*

Proof. We note that after k increases and $i - k$ decreases in the stock price, the resulting price is $u^k d^{i-k}S(t_0)$, for $k = 0, \ldots, i$. But this can occur in $\binom{i}{k}$ different ways, each with probability $p^k(1 - p)^{i-k}$.

Hence

$$P\left(S(t_i) = u^k d^{i-k} S(t_0)\right) = \binom{i}{k} p^k (1-p)^{i-k}.$$

Thus, the expected value of $S(t_i)$, $E(S(t_i))$, is

$$E(S(t_i)) = \sum_{k=0}^{i} \binom{i}{k} (pu)^k ((1-p)d)^{i-k} S(t_0) = S(t_0) \left(pu + (1-p)d\right)^i.$$

But we have seen that $pu + (1-p)d = e^{i^{(\infty)}\delta t}$. Therefore $E(S(t_i)) = S(t_0)e^{i^{(\infty)}i\delta t}$. $\square$

Let $S(t)$ be the price of the stock at time t, and consider times $0 = t_0 < t_1 < \cdots < t_n = T$, with $\Delta t = t_{i+1} - t_i = T/n$ for $i = 0, 1, \ldots, n-1$. If at time t_i the price of the stock is $S(t_i)$, then at time t_{i+1} the price of the stock is $uS(t_i)$ with probability p and $dS(t_i)$ with probability $1-p$. Here $u = e^{\sigma\sqrt{\Delta t}}$ and $d = e^{-\sigma\sqrt{\Delta t}}$, and so from (11.5), we have

$$p = \frac{e^{i^{(\infty)}\Delta t} - d}{u - d} = \frac{e^{i^{(\infty)}\Delta t} - e^{-\sigma\sqrt{\Delta t}}}{e^{\sigma\sqrt{\Delta t}} - e^{-\sigma\sqrt{\Delta t}}}.$$

Note that p and Δt are functions of n because $\Delta t = T/n$. By making n large enough, we can ensure that

$$e^{-\sigma\sqrt{\Delta t}} < e^{i^{(\infty)}\Delta t} < e^{\sigma\sqrt{\Delta t}},$$

that is, $d < e^{i^{(\infty)}\Delta t} < u$, which guarantees that $0 < p < 1$.

Theorem 11.5. *With these choices of u, d, and p, the random variables $S(t_n)/S(0)$, $n = 1, 2, \ldots$, converge in distribution to a lognormal random variable with parameters $\left(i^{(\infty)} - \frac{1}{2}\sigma^2\right)T$ and $\sigma^2 T$.*

Proof. For every n we define

$$X_i = \begin{cases} 1 \text{ if } S(t_i) = uS(t_{i-1}), \\ 0 \text{ if } S(t_i) = dS(t_{i-1}), \end{cases}$$

for $i = 1, 2, \ldots, n$. Then the random variables X_i, $i = 1, 2, \ldots, n$, are independent binomial random variables with parameters 1 and p. If we let $X = \sum_{i=1}^{n} X_i$, then X is a binomial random variable with parameters n and p. It follows that

$$S(t_n) = S(n\Delta t) = S(t_0)u^X d^{n-X} = S(t_0)d^n \left(\frac{u}{d}\right)^X,$$

or

$$\frac{S(t_n)}{S(t_0)} = d^n \left(\frac{u}{d}\right)^X.$$

Thus,

$$\ln\left(\frac{S(t_n)}{S(t_0)}\right) = n\ln d + X\ln\left(\frac{u}{d}\right).$$

But,

$$\ln d = \ln\left(e^{-\sigma\sqrt{\Delta t}}\right) = -\sigma\sqrt{\Delta t},$$

and

$$\ln\left(\frac{u}{d}\right) = \ln\left(\frac{e^{\sigma\sqrt{\Delta t}}}{e^{-\sigma\sqrt{\Delta t}}}\right) = 2\sigma\sqrt{\Delta t},$$

so

$$\ln\left(\frac{S(t_n)}{S(t_0)}\right) = -n\sigma\sqrt{\Delta t} + 2X\sigma\sqrt{\Delta t}.$$

We rewrite this as

$$\ln\left(\frac{S(t_n)}{S(t_0)}\right) = (2p-1)n\sigma\sqrt{\Delta t} + 2\left(X-np\right)\sigma\sqrt{\Delta t}$$

$$= (2p-1)n\sigma\sqrt{\Delta t} + \frac{(X-np)}{\sqrt{np\left(1-p\right)}}2\sqrt{np\left(1-p\right)}\sigma\sqrt{\Delta t}.$$

Now,

$$p = \frac{e^{i^{(\infty)}\Delta t} - d}{u - d}$$

$$= \frac{e^{i^{(\infty)}\Delta t} - e^{-\sigma\sqrt{\Delta t}}}{e^{\sigma\sqrt{\Delta t}} - e^{-\sigma\sqrt{\Delta t}}}$$

$$= \frac{\sigma\sqrt{\Delta t} + i^{(\infty)}\Delta t - \frac{1}{2}\sigma^2\Delta t}{2\sigma\sqrt{\Delta t}} + \text{terms of order } \Delta t$$

$$= \frac{1}{2} + \frac{i^{(\infty)} - \frac{1}{2}\sigma^2}{2\sigma}\sqrt{\Delta t} + \text{terms of order } \Delta t,$$

so

$$(2p-1)n\sigma\sqrt{\Delta t} = \left(\frac{i^{(\infty)} - \frac{1}{2}\sigma^2}{\sigma}\sqrt{\Delta t} + \text{terms of order } \Delta t\right)\frac{\sigma T}{\sqrt{\Delta t}},$$

giving

$$\lim_{n\to\infty}(2p-1)n\sigma\sqrt{\Delta t} = \left(i^{(\infty)} - \frac{1}{2}\sigma^2\right)T.$$

Also, we have

$$\lim_{n\to\infty}2\sqrt{np\left(1-p\right)}\sigma\sqrt{\Delta t} = \lim_{n\to\infty}2\sqrt{p\left(1-p\right)}\sigma\sqrt{T}$$

$$= \sigma\sqrt{T}.$$

From the Binomial Convergence Theorem on p. 273 of Appendix B.4, it follows that

$$Z_n = \frac{(X - np)}{\sqrt{np\,(1 - p)}} \xrightarrow{d} Z,$$

the standard normal random variable. It follows immediately that

$$\ln\left(\frac{S(t_n)}{S(t_0)}\right) \xrightarrow{d} \left(i^{(\infty)} - \frac{1}{2}\sigma^2\right) T + \sigma\sqrt{T}Z,$$

a normal random variable with mean $\left(i^{(\infty)} - \frac{1}{2}\sigma^2\right) T$ and variance $\sigma^2 T$.   □

11.7 Pricing of Options

We now discuss option pricing using binomial trees. We consider a standard binomial tree with initial stock price $S(t_0)$ and initial call option price $C(t_0)$. At time t_1 the stock price is $S_U(t_1)$ or $S_D(t_1)$ with $S_D(t_1) < S_U(t_1)$. These are the only two possibilities. For a one-step tree, we have $C_U(t_1)$ and $C_D(t_1)$ as the call option prices at time t_1. We construct a portfolio consisting of Δ shares of the stock and cash in a risk-free investment with a future value of B, which costs $\Delta S(t_0) + Be^{-i^{(\infty)}\delta_t}$, where $i^{(\infty)}$ is the continuously compounded risk-free rate of return and $\delta_t = t_1 - t_0$.

At time t_1 the value of our portfolio is $\Delta S_U(t_1) + B$ if the stock price is $S_U(t_1)$ and $\Delta S_D(t_1) + B$ if the stock price is $S_D(t_1)$.

Suppose that we want the value of the portfolio to match the option price in both cases, that is,

$$\Delta S_U(t_1) + B = C_U(t_1),$$
$$\Delta S_D(t_1) + B = C_D(t_1).$$

Solving for Δ and B gives

$$\Delta = \frac{C_U(t_1) - C_D(t_1)}{S_U(t_1) - S_D(t_1)},$$

and

$$B = C_U(t_1) - \frac{C_U(t_1) - C_D(t_1)}{S_U(t_1) - S_D(t_1)}S_U(t_1) = \frac{C_D(t_1)S_U(t_1) - C_U(t_1)S_D(t_1)}{S_U(t_1) - S_D(t_1)}.$$

Because the value of our portfolio at time t_1 agrees with the value of the call option at time t_1, we need $C(t_0) = \Delta S(t_0) + Be^{-i^{(\infty)}\delta_t}$. Thus,

$$C(t_0) = \frac{C_U(t_1) - C_D(t_1)}{S_U(t_1) - S_D(t_1)}S(t_0) + \left(\frac{C_D(t_1)S_U(t_1) - C_U(t_1)S_D(t_1)}{S_U(t_1) - S_D(t_1)}\right) e^{-i^{(\infty)}\delta_t}.$$

So to hedge we buy $(C_U(t_1) - C_D(t_1))/(S_U(t_1) - S_D(t_1))$ shares of the stock and purchase a risk-free investment with face value B.

In this case, it can be shown that

$$C(t_0)e^{i^{(\infty)}\delta_t} = pC_U(t_1) + (1 - p)C_D(t_1),$$

with

$$p = \frac{S(t_0)e^{i^{(\infty)}\delta_t} - S_D(\bar{t}_1)}{S_U(t_1) - S_D(t_1)}.$$

Thus, if

$$0 < \frac{S(t_0)e^{i^{(\infty)}\delta_t} - S_D(t_1)}{S_U(t_1) - S_D(t_1)} < 1,$$

then the quantity $C(t_0)e^{i^{(\infty)}\delta_t}$ can be thought of as the expected value of the call option price at time t_1. Note that this is equivalent to $d < e^{i^{(\infty)}\delta_t} < u$ in the case of a standard binomial tree.

We may also write this as

$$C(t_0) = [pC_U(t_1) + (1 - p)C_D(t_1)]\, e^{-i^{(\infty)}\delta_t}, \tag{11.6}$$

which can be thought of as the expected value of the call option price at time t_1, discounted to time t_0.

Note that if $S_D(t_1) < S_U(t_1)$ and

$$\frac{1}{\delta_t}\ln\left(\frac{S_D(t_1)}{S(t_0)}\right) < i^{(\infty)} < \frac{1}{\delta_t}\ln\left(\frac{S_U(t_1)}{S(t_0)}\right),$$

then $0 < p < 1$ holds, that is,

$$0 < \frac{S(t_0)e^{i^{(\infty)}\delta_t} - S_D(t_1)}{S_U(t_1) - S_D(t_1)} < 1.$$

If we do not have a standard binomial tree, that is, the up and down ratios u and d are not constant, then the transition probabilities may differ.

We now generalize the one-step model to any number of steps. Let us suppose that we have an n-step standard binomial tree, starting at time t_0, with t_n being the expiration time of the corresponding call option. Suppose that we also know the stock price at each node of the tree and that we know the exercise price of the corresponding call option. We may then calculate the fair price of the call option, as given by (11.6), at every node of the tree.

Consider the tree given previously in Table 11.3 on p. 216, which we reproduce here.

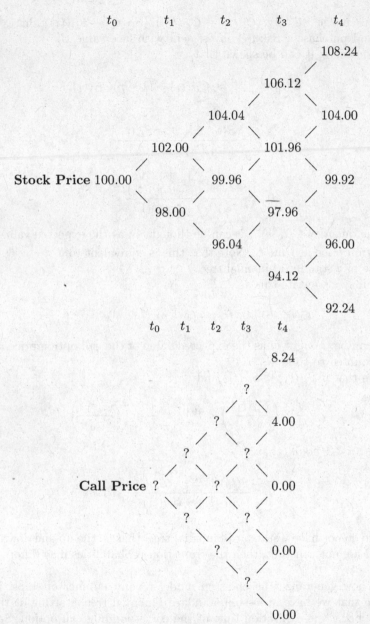

Note that the call prices at time t_4 (expiration) are determined uniquely by the corresponding stock prices at that time. That is, the call prices at time t_4 are $S(t_4) - 100$ if $S(t_4) > 100$ and 0 if $S(t_4) \leq 100$.

Assuming an exercise price of 100, we use "backward induction" to fill in the missing values. Consider the call prices at time t_3 with $i^{(\infty)} = 0.05$ and $\delta t = 1/12$. We need

$$C(t_3) = [pC_U(t_4) + (1-p)C_D(t_4)] e^{-i(\infty)\delta_t},$$

where $e^{-i(\infty)\delta_t} = e^{-1/240}$. Also,

$$p = \frac{e^{1/240}S(t_3) - S_D(t_4)}{S_U(t_4) - S_D(t_4)}.$$

We next note that because we have a standard binomial tree, this value of p is the same throughout the tree. In this case, $p = 0.6044$.

At time t_3 we have the following for the top branch: $C_U(t_4) = 8.24$ and $C_D(t_4) = 4.00$. Hence

$$\begin{aligned}
C(t_3) &= [pC_U(t_4) + (1-p)C_D(t_4)] e^{-i(\infty)\delta_t} \\
&= [0.6044(8.24) + 0.3956(4.00)] e^{-1/240} \\
&= 6.54.
\end{aligned}$$

Similarly, for the second branch from the top, $S_U(t_4) = 104.00$, $S_D(t_4) = 99.92$, $C_U(t_4) = 4.00$, and $C_D(t_4) = 0.00$. This leads to $C(t_3) = 2.41$. The bottom two branches have $C(t_3) = 0$. Similarly, at time t_2, for the top branch we have $S_U(t_3) = 106.12$, $S_D(t_3) = 101.96$, $C_U(t_2) = 6.54$, and $C_D(t_2) = 2.41$. This yields $C(t_2) = 4.88$.

Completing the calculations yields the following tree:

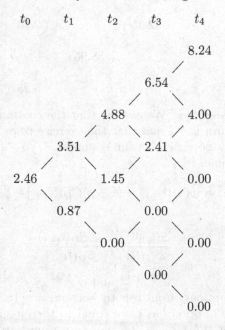

$$t_0 \qquad t_1 \qquad t_2 \qquad t_3 \qquad t_4$$

Notice that the call prices satisfy the relationship $C(t_i) + X \geq S(t_i)$ (call price plus exercise price is greater than or equal to the stock price) at each time t_i. If this were not the case (that is, $C(t_i) + X < S(t_i)$), then an investor could purchase the call, exercise it, and then sell the stock, for a profit of $S(t_i) - (C(t_i) + X) > 0$, a guaranteed profit.

Notice also that several of the call prices are zero. For example, if the stock decreases in price in each of the first two time periods, giving a price of \$96.04, then the highest possible price at expiration (time t_4) is \$99.92, which is less than the exercise price.

A binomial tree that is not a standard binomial tree is called a **flexible tree**. Given the stock price at each node and the exercise price, we may still calculate the fair price of an option[12] if we know the continuously compounded risk-free rate of return.

Consider the following example for the stock of company WILD:

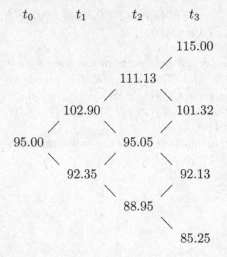

This is a three-step tree. We assume that the continuously compounded risk-free rate of return is 4% and that the exercise price is \$100. As before, $\delta_t = 1/12$, so the option expires in three months.

We use the formulas

$$C(t_i) = [pC_U(t_{i+1}) + (1-p)C_D(t_{i+1})] e^{-i(\infty)\delta_t}$$

and

$$p = \frac{S(t_i)e^{i(\infty)\delta_t} - S_D(t_{i+1})}{S_U(t_{i+1}) - S_D(t_{i+1})}.$$

We calculate $e^{-i(\infty)\delta_t} = e^{-0.04/12}$ and $e^{i(\infty)\delta_t} = e^{0.04/12}$. At time t_3 the values of the call options, from top to bottom, are 15.00, 1.32, 0.00, and 0.00 because the call option has price equal to the larger of $S(t_3) - X$ and 0 at expiration. For times t_0, t_1, t_2, and t_3 we must recalculate p at each

---

[12] The fair price of an option is the price at which no arbitrage opportunities exist.

branch. For example, at time t_2 for the upper branch we have $S(t_2) = 111.13$, $S_U(t_3) = 115.00$, and $S_D(t_3) = 101.32$. Then

$$p = \frac{e^{0.04/12}(111.13) - 101.32}{115.00 - 101.32} = 0.7442,$$

and

$$C(t_2) = [0.7442(15.00) + 0.2558(1.32)]e^{-0.04/12} = 11.46.$$

In Problem 11.26 on p. 241 you are asked to compute the call option prices for this tree.

11.8 The Black-Scholes Option Pricing Model

In Theorem 11.5 on p. 218 we investigated a particular binomial model of stock prices, which in the limit converges to a Geometric Brownian Motion Model, where at each time t the ratio $S(t)/S(0)$ is a lognormal random variable with parameters $\left(i^{(\infty)} - \frac{1}{2}\sigma^2\right)T$ and $\sigma^2 T$. In this section we derive the Black-Scholes formula, which gives the fair (arbitrage-free) price, $C(0)$, of a European call option on a stock using this Geometric Brownian Motion Model.

We point out that, by definition, the natural logarithm of a lognormal random variable is a normal random variable. However, if this normal random variable has expected value μ and variance σ^2, then the lognormal random variable has expected value $e^{\mu+\sigma^2/2}$. These considerations lead to the lognormal random variable (with parameters $\left(i^{(\infty)} - \frac{1}{2}\sigma^2\right)T$ and $\sigma^2 T$) as the appropriate one to use. We can show that, under these conditions, $S(t)$ has a lognormal distribution with $E(S(t)) = e^{i^{(\infty)}t}S(0)$. (See Problems 11.29 and 11.30.) We now find the fair call option price for a stock, the price of which has this distribution.

In the following the stock price $S(t)$ and the volatility $\sigma > 0$ are as in Theorem 11.5 on p. 218. The exercise price is X, the time to expiration is T in years, and the continuously compounded risk-free rate of return is $i^{(\infty)}$.

Derivation of the Black-Scholes Formula

Recall from (11.1) on p. 194, that at time T the price of the stock, $S(T)$, the price of the call option, $C(T)$, and the exercise price, X are related by

$$C(T) = \begin{cases} 0 & \text{if } S(T) \leq X, \\ S(T) - X & \text{if } S(T) > X, \end{cases}$$

so[13]

$$C(T) = [S(T) - X]^+.$$

---

[13] The function $[x]^+$ is defined to be x if $x > 0$, and 0 if $x \leq 0$.

In order to find the fair price of the option, $C(0)$, at time[14] $t = 0$ we discount the expected value of the call option on the expiration date, $E(C(T))$, using the continuously compounded risk-free rate of return, so

$$C(0) = e^{-i^{(\infty)}T}E(C(T)) = e^{-i^{(\infty)}T}E\left(\left[S(T) - X\right]^+\right).$$

Now,

$$S(T) = S(0)\frac{S(T)}{S(0)} = S(0)e^{\ln(S(T)/S(0))} = S(0)e^W,$$

where $W = \ln(S(T)/S(0))$, and thus W is a normal random variable with mean $\left(i^{(\infty)} - \frac{1}{2}\sigma^2\right)T$, variance $\sigma^2 T$, and density function

$$f(w) = \frac{1}{\sqrt{2\pi\sigma^2 T}}e^{-\frac{1}{2}\left(w - \left(i^{(\infty)} - \frac{1}{2}\sigma^2\right)T\right)^2/(\sigma^2 T)},$$

for $-\infty < w < \infty$.

Thus, we have

$$C(0) = e^{-i^{(\infty)}T}E\left(\left[S(T) - X\right]^+\right) = e^{-i^{(\infty)}T}E\left(\left[S(0)e^W - X\right]^+\right),$$

so all that remains is to calculate $E\left(\left[S(0)e^W - X\right]^+\right)$, which we do now.

We have

$$E\left(\left[S(0)e^W - X\right]^+\right)$$

$$= \frac{1}{\sqrt{2\pi\sigma^2 T}}\int_{-\infty}^{\infty}\left[S(0)e^W - X\right]^+ e^{-\frac{1}{2}\left(w - \left(i^{(\infty)} - \frac{1}{2}\sigma^2\right)T\right)^2/(\sigma^2 T)}\,dw.$$

However, $S(0)e^W - X \geq 0$ if and only if $W \geq \ln\left(X/(S(0))\right)$, so we have

$$E\left(\left[S(0)e^W - X\right]^+\right)$$

$$= \frac{1}{\sqrt{2\pi\sigma^2 T}}\int_{\ln(X/S(0))}^{\infty}\left(S(0)e^w - X\right)e^{-\frac{1}{2}\left(w - \left(i^{(\infty)} - \frac{1}{2}\sigma^2\right)T\right)^2/(\sigma^2 T)}\,dw.$$

This integral is the difference between two integrals, I_1 and I_2, that is,

$$E\left(\left[S(0)e^W - X\right]^+\right) = I_1 - I_2,$$

where

---

[14] We may always assume that $t = 0$ (the present time is 0) because it is only the time **remaining** until expiration, T, that matters (in addition to $S(0)$, X, $i^{(\infty)}$, and σ^2) in calculating the fair price of an option.

$$I_1 = \frac{1}{\sqrt{2\pi\sigma^2 T}} \int_{\ln(X/S(0))}^{\infty} S(0)e^w e^{-\frac{1}{2}\left(w-\left(i^{(\infty)}-\frac{1}{2}\sigma^2\right)T\right)^2/(\sigma^2 T)}\, dw$$

$$= \frac{S(0)}{\sqrt{2\pi\sigma^2 T}} \int_{\ln(X/S(0))}^{\infty} e^{w-\frac{1}{2}\left(w-\left(i^{(\infty)}-\frac{1}{2}\sigma^2\right)T\right)^2/(\sigma^2 T)}\, dw,$$

and[15]

$$I_2 = \frac{1}{\sqrt{2\pi\sigma^2 T}} \int_{\ln(X/S(0))}^{\infty} X e^{-\frac{1}{2}\left(w-\left(i^{(\infty)}-\frac{1}{2}\sigma^2\right)T\right)^2/(\sigma^2 T)}\, dw,$$

which we evaluate in turn.

In order to evaluate I_1 we set

$$u = \frac{w - \left(i^{(\infty)} + \frac{1}{2}\sigma^2\right)T}{\sigma\sqrt{T}},$$

so that

$$w - \frac{1}{2}\frac{\left(w - \left(i^{(\infty)} - \frac{1}{2}\sigma^2\right)T\right)^2}{\sigma^2 T}$$

$$= -\frac{1}{2}\frac{-2\sigma^2 Tw + \left(w^2 - 2\left(i^{(\infty)} - \frac{1}{2}\sigma^2\right)Tw + \left(i^{(\infty)} - \frac{1}{2}\sigma^2\right)^2 T^2\right)}{\sigma^2 T}$$

$$= -\frac{1}{2}\frac{\left(w^2 - 2\left(i^{(\infty)} + \frac{1}{2}\sigma^2\right)Tw + \left(i^{(\infty)} + \frac{1}{2}\sigma^2\right)^2 T^2\right) - 2i^{(\infty)}\sigma^2 T^2}{\sigma^2 T}$$

$$= -\frac{1}{2}u^2 + i^{(\infty)}T,$$

giving

$$e^{w-\frac{1}{2}\left(w-\left(i^{(\infty)}-\frac{1}{2}\sigma^2\right)T\right)^2/(\sigma^2 T)} = e^{-\frac{1}{2}u^2} e^{i^{(\infty)}T}.$$

Substituting this into I_1, we find that

$$I_1 = S(0)e^{i^{(\infty)}T}\frac{1}{\sqrt{2\pi}}\int_{-\delta}^{\infty} e^{-\frac{1}{2}u^2}\, du$$

$$= S(0)e^{i^{(\infty)}T}\left(1 - \Phi\left(-\delta\right)\right),$$

where δ and $\Phi(x)$ are given by

$$\delta = \frac{\left(i^{(\infty)} + \frac{1}{2}\sigma^2\right)T - \ln\left(X/S(0)\right)}{\sigma\sqrt{T}},$$

---

[15] Note that I_2 is equal to the exercise price, X, times the probability that the option is in the money at time T.

and

$$\Phi(x) = \frac{1}{\sqrt{2\pi}} \int_{-\infty}^{x} e^{-t^2/2} \, dt.$$

However (see Problem 11.32),

$$\Phi(\delta) = 1 - \Phi(-\delta),$$

so

$$I_1 = S(0)e^{i^{(\infty)}T}\Phi(\delta).$$

We now turn to I_2. Substituting

$$v = \frac{w - \left(i^{(\infty)} - \frac{1}{2}\sigma^2\right)T}{\sigma\sqrt{T}}$$

into I_2 we have

$$
\begin{aligned}
I_2 &= \frac{X}{\sqrt{2\pi}} \int_{-\delta+\sigma\sqrt{T}}^{\infty} e^{-\frac{1}{2}v^2} \, dv \\
&= X\left(1 - \Phi\left(-\delta + \sigma\sqrt{T}\right)\right) \\
&= X\Phi\left(\delta - \sigma\sqrt{T}\right).
\end{aligned}
$$

Finally, we find that

$$E\left(\left[S(0)e^W - X\right]^+\right) = e^{i^{(\infty)}T}S(0)\Phi(\delta) - X\Phi\left(\delta - \sigma\sqrt{T}\right),$$

which, when substituted into

$$C(0) = e^{-i^{(\infty)}T}E\left(\left[S(0)e^W - X\right]^+\right),$$

leads to the famous **Black-Scholes formula**:

$$C(0) = S(0)\Phi(\delta) - Xe^{-i^{(\infty)}T}\Phi\left(\delta - \sigma\sqrt{T}\right), \tag{11.7}$$

where δ and $\Phi(x)$ are given by

$$\delta = \frac{\left(i^{(\infty)} + \frac{1}{2}\sigma^2\right)T - \ln\left(X/S(0)\right)}{\sigma\sqrt{T}}, \tag{11.8}$$

and

$$\Phi(x) = \frac{1}{\sqrt{2\pi}} \int_{-\infty}^{x} e^{-t^2/2} \, dt. \tag{11.9}$$

Significance of the Black-Scholes Formula

The Black-Scholes formula (11.7) gives the price of a European call option assuming that the price of a stock follows a Geometric Brownian Motion. The theory leading to this result is given in [3] and is still of great importance. In 1997, Myron Scholes and Robert Merton shared the Nobel prize in Economics for its development.[16]

The Black-Scholes formula (11.7) is a major result for several reasons. It is one of the results that ushered in the modern theory of finance, which is based in large part on mathematical reasoning. There are five quantities in the Black-Scholes pricing formula, of which four are observable, so there is only one quantity, σ, that needs to be estimated.

Black-Scholes Assumptions

This derivation of the Black-Scholes formula (11.7) depends implicitly on the following two assumptions.

1. *The price of the stock follows a Geometric Brownian Motion.*
   As a result of this assumption we have the following consequences.
   - The continuously compounded risk-free rate of return, $i^{(\infty)}$, is constant. The risk-free rate is used to discount the expected value of the call option on the expiration date. The risk-free rate also appears in the parameter $\left(i^{(\infty)} - \frac{1}{2}\sigma^2\right) T$ of the Geometric Brownian Motion Model, so if it is not constant, then the assumption that the price of the stock follows a Geometric Brownian Motion is violated.
   - The volatility, σ, is constant. If this constant volatility assumption is not made, then changes in the value of the option may occur even if the stock price is unchanged. This would be a violation of the model.
   - The stock price follows a continuous path over time. Otherwise there would be jumps in the price of the stock, which violates this model.
   - There are no dividends during the life of the stock. Dividends are often associated with jumps in the price of the stock. Such jumps violate the assumption that the price of the stock follows a Geometric Brownian Motion.[17]
   - At time t the ratio $S(t)/S(0)$ has a lognormal distribution.
2. *There are no arbitrage opportunities.*

These assumptions are generally not satisfied in the real world. For example, jumps, occasionally large, in the stock price do occur. However, the significance of this result, which provides a mathematical framework for option pricing, should not be underestimated.

---

[16] Fischer Black was deceased when the Nobel Prize was awarded. Robert Merton, in [19], had developed a similar theory on options during the same time period.

[17] However, depending on the type of dividend, alternative models are available.

Comments on the Black-Scholes Formula

- The Black-Scholes formula (11.7) allows us to find an exact solution for the fair price of the call option, which is one of the surprising properties of this model.
- The Black-Scholes formula (11.7) consists of two parts, $S(0)\Phi(\delta)$ and $Xe^{-i^{(\infty)}T}\Phi\left(\delta - \sigma\sqrt{T}\right)$. In Problem 11.35 you are asked to show that

$$\Phi(\delta) = \frac{\partial C(0)}{\partial S(0)}.$$

In Problem 11.36 you are asked to show that $\Phi\left(\delta - \sigma\sqrt{T}\right)$ is the probability that the option is in the money at expiration, that is, $S(T) > X$. Thus, the Black-Scholes formula can be interpreted as: $C(0)$ is equal to the present stock price $S(0)$ times the rate of change of $C(0)$ with respect to $S(0)$ plus the present value of a cash flow equal to X if the option is in the money at expiration $(S(T) > X)$ and equal to 0 otherwise.
- The only unobservable quantity, σ, is often estimated from historical data. Another method uses what is called *implied volatility*. If we know $S(0)$, $C(0)$, X, T, and $i^{(\infty)}$, then we can estimate σ from the Black-Scholes formula (11.7) because it is the only missing quantity. This requires numerical methods because there is no analytic solution of (11.7) for σ in terms of the remaining quantities.
- Using the Put-Call Parity Theorem on p. 209, we can derive the Black-Scholes price for a European put option, namely

$$P(0) = -S(0)\Phi(-\delta) + e^{-i^{(\infty)}T}\Phi\left(-\delta + \sigma\sqrt{T}\right).$$

(See Problem 11.33.)
- Formulas exist for European options on stocks with dividends, and for some American call options. At the present time there is no known analytic formula for the value of an American put.
- We have only introduced you to the subject of Black-Scholes. There are excellent books and articles—such as [5] and [7]—which treat this subject in greater depth.
- To use the Black-Scholes formula (11.7) we must estimate

$$\Phi(x) = \frac{1}{\sqrt{2\pi}} \int_{-\infty}^{x} e^{-t^2/2}\, dt,$$

for different values of x. Immediately following these comments, we digress to show various ways of doing this. If this is not new to you, please proceed to Example 11.4 on p. 235.

Digression on Calculating $\Phi(x)$

To apply the Black-Scholes formula (11.7) we must be able to estimate

$$\Phi(x) = \frac{1}{\sqrt{2\pi}} \int_{-\infty}^{x} e^{-t^2/2} \, dt,$$

for different values of x. In this section we discuss four different ways of doing this, depending on the resources available. This is done in the context of estimating $\Phi(0.1584)$, $\Phi(-0.1245)$, and $98\Phi(0.1584) - 100e^{-0.025}\Phi(-0.1245)$, which are used in Example 11.4.

1. **Tables**

   Here we use a look-up table, such as Table B.1 on p. 268 (which has values from 0.00 to 2.99, in steps of 0.01, to 4 decimal places) or Table 26.1 on p. 966 of [1] (which has values from 0.00 to 2.00, in steps of 0.02, to 15 decimal places).

   The advantage of this method is that all we have to do is to turn to either Table B.1 or Table 26.1 in [1] and look up the value of $\Phi(0.1584)$. The disadvantage is that 0.1584 is not one of the entries in either table. So what do we do? We round 0.1584 to the nearest entry in the table. We now perform the calculations using both tables and denote these estimates by $\Phi_{T_1}(x)$ and $\Phi_{T_2}(x)$.

   (a) The nearest entry in Table B.1 to 0.1584 is 0.16, so we estimate that

   $$\Phi_{T_1}(0.1584) \approx \Phi(0.16) = 0.5636.$$

   There is a further disadvantage when we try to estimate $\Phi(-0.1245)$ because negative values are not in this table (nor in Table 26.1 of [1]). In this case we need to use the identity (see Problem 11.32)

   $$\Phi(-x) = 1 - \Phi(x),$$

   so that

   $$\Phi(-0.1245) = 1 - \Phi(0.1245).$$

   Again 0.1245 is not one of the entries in Table B.1, so we round 0.1245 to the nearest entry in this table, which is 0.12, and estimate that

   $$\Phi_{T_1}(0.1245) \approx \Phi(0.12) = 0.5478,$$

   so

   $$\Phi_{T_1}(-0.1245) \approx 1 - 0.5478 = 0.4522.$$

   Thus,

   $$98\Phi(0.1584) - 100e^{-0.025}\Phi(-0.1245) \approx 11.1293.$$

(b) The nearest entry in Table 26.1 of [1] to 0.1584 is 0.16, so we estimate that

$$\Phi_{T_2}(0.1584) \approx \Phi(0.16) = 0.563559462891433.$$

To estimate $\Phi(-0.1245)$ we again use the identity

$$\Phi(-0.1245) = 1 - \Phi(0.1245).$$

Once more 0.1245 is not one of the entries in Table 26.1, so we round 0.1245 to the nearest entry on this table, which is 0.12, and estimate that

$$\Phi_{T_2}(0.1245) \approx \Phi(0.12) = 0.547758426020584,$$

so

$$\Phi_{T_2}(-0.1245) \approx 1 - 0.547758426020584 = 0.452241573979416.$$

Thus,

$$98\Phi(0.1584) - 100e^{-0.025}\Phi(-0.1245) \approx 11.121258390018500.$$

2. **Linear Interpolation**
    Here we use one of the tables mentioned in the first part, together with linear interpolation described by (6.7) on p. 95. We now perform the calculations using both tables and denote these estimates by $\Phi_{LI_1}(x)$ and $\Phi_{LI_2}(x)$.
    (a) From Table B.1, we see that the nearest entries on either side of 0.1584 are 0.15 and 0.16, with $\Phi(0.15) = 0.5596$ and $\Phi(0.16) = 0.5636$. Using (6.7) with $l = 0.1584$, $a = 0.15$, and $b = 0.16$, gives

$$\Phi_{LI_1}(0.1584) = \frac{0.16 - 0.1584}{0.16 - 0.15}\Phi(0.15) + \frac{0.1584 - 0.15}{0.16 - 0.15}\Phi(0.16)$$
$$= 0.16(0.5596) + 0.84(0.5636)$$
$$= 0.56296.$$

In the same way

$$\Phi_{LI_1}(-0.1245) = 1 - \Phi_{LI_1}(0.1245)$$
$$= 1 - \left(\frac{0.13 - 0.1245}{0.13 - 0.12}\Phi(0.12) + \frac{0.1245 - 0.12}{0.13 - 0.12}\Phi(0.13)\right)$$
$$= 1 - 0.55(0.5478) - 0.45(0.5517)$$
$$= 0.450445.$$

Thus,

$$98\Phi(0.1584) - 100e^{-0.025}\Phi(-0.1245) \approx 11.2377.$$

(b) From Table 26.1 of [1], we see that the nearest entries on either side of 0.1584 are 0.14 nd 0.16, with $\Phi(0.14) = 0.555670004805907$ and $\Phi(0.16) = 0.563559462891433$. Using (6.7) with $l = 0.1584$, $a = 0.14$, and $b = 0.16$, gives

$$\Phi_{LI_2}(0.1584) = \frac{0.16 - 0.1584}{0.16 - 0.14}\Phi(0.15) + \frac{0.1584 - 0.14}{0.16 - 0.14}\Phi(0.16)$$
$$= 0.08(0.555670004805907) + 0.92(0.563559462891433)$$
$$= 0.562928306244591.$$

In the same way

$$\Phi_{LI_2}(-0.1245) = 1 - \Phi_{LI_2}(0.1245)$$
$$= 1 - \left(\frac{0.14 - 0.1245}{0.14 - 0.12}\Phi(0.12) + \frac{0.1245 - 0.12}{0.14 - 0.12}\Phi(0.14)\right)$$
$$= 1 - 0.775(0.547758426020584)$$
$$\qquad - 0.225(0.555670004805907)$$
$$= 0.450461468752718.$$

Thus,

$$98\Phi(0.1584) - 100e^{-0.025}\Phi(-0.1245) \approx 11.233020465833200.$$

Because $\Phi(x)$ is positive for all x and concave down for $x > 0$ (see Problem 11.32), we can use the comments following Theorem 6.4 on p. 96 to conclude that $\Phi_{LI_2}(0.1584)$ is an underestimate for $\Phi_{LI_2}(0.1584)$, that is,

$$\Phi_{LI_2}(0.1584) = 0.562928306244591 < \Phi(0.1584).$$

Also, because $\Phi(x)$ is concave up for $x < 0$, the value of $\Phi_{LI_2}(-0.1245)$ is an overestimate for $\Phi(-0.1245)$, that is,

$$\Phi_{LI_2}(-0.1245) = 0.450461468752718 > \Phi(-0.1245).$$

Finally, because

$$\Phi_{LI_2}(0.1584) < \Phi(0.1584) \text{ and } -\Phi_{LI_2}(-0.1245) < -\Phi(-0.1245),$$

we see that

$$98\Phi_{LI_2}(0.1584) - 100e^{-0.025}\Phi_{LI_2}(-0.1245)$$
$$< 98\Phi(0.1584) - 100e^{-0.025}\Phi(-0.1245),$$

so 11.233020465833200 is an underestimate, that is,

$$11.233020465833200 < 98\Phi(0.1584) - 100e^{-0.025}\Phi(-0.1245).$$

3. **Numerical Approximations**

Here we use a numerical approximation. There are various formulas used to approximate the value of $\Phi(x)$ (see p. 932 of [1]). We present two, which we denote by $\Phi_{NA_1}(x)$ and $\Phi_{NA_2}(x)$.

(a) The first approximation is

$$\Phi_{NA_1}(x) = 1 - \frac{e^{-x^2/2}}{\sqrt{2\pi}}\left(a_1 t + a_2 t^2 + a_3 t^3\right),$$

where

$$t = \frac{1}{1 + px},$$

with $p = 0.33267$, $a_1 = 0.4361836$, $a_2 = -0.1201676$, and $a_3 = 0.9372980$. This approximation has an error of less than 1.0×10^{-5}. Using this numerical approximation we find that

$$\Phi_{NA_1}(0.1584) = 0.5629382,$$
$$\Phi_{NA_1}(-0.1245) = 0.4503966,$$
$$98\Phi(0.1584) - 100e^{-0.025}\Phi(-0.1245) \approx 11.2403182.$$

(b) The second approximation is

$$\Phi_{NA_2}(x) = 1 - \frac{e^{-x^2/2}}{\sqrt{2\pi}}\left(b_1 t + b_2 t^2 + b_3 t^3 + b_4 t^4 + b_5 t^5\right),$$

where

$$t = \frac{1}{1 + px},$$

with $p = 0.2316419$, $b_1 = 0.319381530$, $b_2 = -0.356563782$, $b_3 = 1.781477937$, $b_4 = -1.821255978$, and $b_5 = 1.330274429$. This approximation has an error of less than 7.5×10^{-8},

Using this numerical approximation we find that

$$\Phi_{NA_2}(0.1584) = 0.56292921,$$
$$\Phi_{NA_2}(-0.1245) = 0.45045863,$$
$$98\Phi_{NA_2}(0.1584) - 100e^{-0.025}\Phi_{NA_2}(-0.1245) = 11.23338582.$$

4. **Exact**

Here we use an "Exact" result using either a spreadsheet or a sophisticated scientific or financial calculator. We denote this value by $\Phi_E(x)$.

In Microsoft® Excel, the command $NORMSDIST(x)$ calculates $\Phi(x)$. Using this we find that

$$\Phi_E(0.1584) = 0.56292920,$$
$$\Phi_E(-0.1245) = 0.45045966,$$
$$98\Phi(0.1584) - 100e^{-0.025}\Phi(-0.1245) \approx 11.23328500.$$

Notice that, as expected when compared to the linear interpolation estimates,

$$\Phi_{LI_2}(0.1584) < \Phi_E(0.1584),$$
$$\Phi_{LI_2}(-0.1245) > \Phi_E(-0.1245),$$

and

$$98\Phi_{LI_2}(0.1584) - 100e^{-0.025}\Phi_{LI_2}(-0.1245)$$
$$< 98\Phi_E(0.1584) - 100e^{-0.025}\Phi_E(-0.1245).$$

Table 11.4 compares these approximations, rounded to 4 decimal places. Now that we can estimate $\Phi(x)$ we can use the Black-Scholes formula (11.7).

Table 11.4. Comparison of Different Estimates for $\Phi(x)$

	$\Phi(0.1584)$	$\Phi(-0.1245)$	$98\Phi(0.1584) - 100e^{-0.025}\Phi(-0.1245)$
T_1	0.5636	0.4522	11.1293
T_2	0.5636	0.4522	11.1213
LI_1	0.5630	0.4504	11.2377
LI_2	0.5629	0.4505	11.2330
NA_1	0.5629	0.4504	11.2403
NA_2	0.5629	0.4505	11.2334
E	0.5629	0.4505	11.2333

Example 11.4. If the current price of the stock is \$98, if the exercise price for a European call option on this stock is \$100, if the time to expiration is 0.5 (one-half year), if the continuously compounded risk-free annual rate of return is 5%, and if the volatility is 0.4, then use the Black-Scholes formula (11.7) to find a fair price for the option.

Solution. Here $S(0) = 98$, $X = 100$, $T = 0.5$, $i^{(\infty)} = 0.05$, and $\sigma = 0.4$. From (11.8) we have

$$\delta = \frac{\left(0.05 + \frac{1}{2}(0.16)\right)0.50 - \ln(100/98)}{0.40\sqrt{0.50}} \approx 0.1584,$$

and

$$\delta - \sigma\sqrt{T} \approx -0.1245,$$

so

$$C(0) = 98\Phi(0.1584) - 100e^{-0.025}\Phi(-0.1245).$$

These are exactly the values we've used to estimate Φ in Table 11.4. Thus, the values we find for $\Phi(0.1584)$ and $\Phi(-0.1245)$, which determine the value of $C(0)$, depend on which of the estimates in Table 11.4 we use.

If we use Table B.1, then we find that $C(0) \approx 11.13$, so the option should cost approximately \$11.13, whereas if we use the "exact" calculation, then we find that $C(0) \approx 11.23$, so the option should cost approximately \$11.23. $\triangle$

Dependence of the Black-Scholes Quantities

There are five quantities used to calculate the Black-Scholes option price $C(0)$, namely $S(0)$, X, T, $i^{(\infty)}$, and σ, so we can think of $C(0)$ as a function of these quantities. There are several relationships between these five quantities and their partial derivatives of $C(0)$. We have already seen that $\Phi(\delta) = \partial C(0)/\partial S(0)$, the rate of change of $C(0)$ with respect to $S(0)$. This partial derivative is called *Delta*, and is denoted by Δ. A list of some of these quantities, called the "Greeks", with their values in terms of the Black-Scholes quantities follows.

- **Delta:**

$$\Delta = \frac{\partial C(0)}{\partial S(0)},$$

measures the sensitivity of the change in price, $C(0)$, to the change in the current price, $S(0)$.

If $C_{dS}(0)$ is the fair price after $S(0)$ has changed in price by dS, then from local linearity,

$$C_{dS}(0) - C(0) \approx dS\frac{\partial C(0)}{\partial S(0)} = dS\Delta.$$

Thus, an increase of \$1 in $S(0)$ leads to a change of approximately \$$\Delta$ in $C(0)$. If we differentiate $C(0)$, given by (11.7), partially with respect to $S(0)$, we find that (see Problem 11.35)

$$\Delta = \frac{\partial C(0)}{\partial S(0)} = \Phi(\delta).$$

Because $\Phi(\delta) > 0$, we have $\partial C(0)/\partial S(0) > 0$, so $C(0)$ increases with $S(0)$, assuming that all other quantities remain constant. Thus, the change of approximately \$$\Delta$ in $C(0)$ is an increase in $C(0)$ by \$$\Delta$, as opposed to a decrease.

- **Gamma:**

$$\Gamma = \frac{\partial \Delta}{\partial S(0)} = \frac{\partial^2 C(0)}{\partial S(0)^2},$$

measures the sensitivity of the change in Δ, to the change in the current price, $S(0)$.

If we differentiate Δ partially with respect to $S(0)$, then we find that (see Problem 11.37)

$$\Gamma = \frac{\partial \Delta}{\partial S(0)} = \frac{\partial^2 C(0)}{\partial S(0)^2} = \frac{\Phi'(\delta)}{S(0)\sigma\sqrt{T}}.$$

Because $\Phi'(\delta)/(S(0)\sigma\sqrt{T}) > 0$, we have $\partial^2 C(0)/\partial S(0)^2 > 0$, so $C(0)$ is concave up when plotted against $S(0)$, assuming that all other quantities remain constant.

- **Theta:**

$$\Theta = \frac{\partial C(0)}{\partial T},$$

measures the sensitivity of the change in price, $C(0)$, to the change in the time until the expiration date, T, of the option.[18]
If we differentiate $C(0)$, given by(11.7), partially with respect to T, then we find that (see Problem 11.38)

$$\Theta = \frac{\partial C(0)}{\partial T} = \frac{S(0)\sigma}{2\sqrt{T}}\Phi'(\delta) + i^{(\infty)}Xe^{-i^{(\infty)}T}\Phi(\delta - \sigma\sqrt{T}).$$

Because the right-hand side is positive, we have $\partial C(0)/\partial T > 0$, so $C(0)$ increases with T, assuming that all other quantities remain constant. Thus, a decrease of one day in T, the time until expiration, leads to a decrease of approximately \$$\Theta/365$ in $C(0)$.

- **rho:**

$$\rho = \frac{\partial C(0)}{\partial i^{(\infty)}},$$

measures the sensitivity of the change in price, $C(0)$, to the change in the continuously compounded risk-free rate of return, $i^{(\infty)}$.
If we differentiate $C(0)$, given by(11.7), partially with respect to $i^{(\infty)}$, then we find that (see Problem 11.39)

$$\rho = \frac{\partial C(0)}{\partial i^{(\infty)}} = XTe^{-i^{(\infty)}T}\Phi(\delta - \sigma\sqrt{T}).$$

Because the right-hand side is positive, we have $\partial C(0)/\partial i^{(\infty)} > 0$, so $C(0)$ increases with $i^{(\infty)}$, assuming that all other quantities remain constant. Thus, an increase of 0.01 in $i^{(\infty)}$ leads to an increase of approximately \$$0.01\rho$ in $C(0)$.

- **Vega:**[19]

$$V = \frac{\partial C(0)}{\partial \sigma},$$

measures the sensitivity of the change in price, $C(0)$, to the change in volatility of the stock, σ.
If we differentiate $C(0)$, given by(11.7), partially with respect to σ, then we find that (see Problem 11.40)

$$V = \frac{\partial C(0)}{\partial \sigma} = S(0)\sqrt{T}\Phi'(\delta).$$

---

[18] Theta is sometimes defined by $\Theta = -\partial C(0)/\partial T$.
[19] This partial derivative is not named after a Greek letter, but after the brightest star in the constellation Lyra.

Because the right-hand side is positive, we have $\partial C(0)/\partial \sigma > 0$, so $C(0)$ increases with σ, assuming that all other quantities remain constant. Thus, an increase of 1 in σ leads to an increase of approximately \$$V$ in $C(0)$.

From these equations we obtain the following condition on some of the Greeks:

$$-\Theta + \frac{1}{2}\sigma^2 S^2(0)\Gamma + i^{(\infty)} S(0)\Delta - i^{(\infty)} C(0) = 0.$$

(See Problem 11.47.) If we replace the Greeks with their partial derivatives, then we obtain the partial differential equation

$$-\frac{\partial F}{\partial T} + \frac{1}{2}\sigma^2 S^2(0)\frac{\partial^2 F}{\partial S(0)^2} + i^{(\infty)} S(0)\frac{\partial F}{\partial S(0)} - i^{(\infty)} F = 0, \qquad (11.10)$$

the solution, F, of which gives the fair price of a call option,[20] $F = C(0)$, or put option,[21] $F = P(0)$, in terms of the stock price $S(0)$, the exercise price X, the time to expiration T, the continuously compounded rate of return $i^{(\infty)}$, and the volatility $\sigma > 0$.

You may have wondered why $\partial C(0)/\partial X$ is not among the Greeks. It is because the quantity X is constant during the life of an option. It is specified in the contract when written. Therefore, the sensitivity of $C(0)$ to changes in X is of no practical significance to buyers and sellers of options. However, it can still be evaluated mathematically, and you are asked to do this in Problem 11.41. As a consequence of this, we see that $C(0)$ is a decreasing function of X. Explain why this is not unexpected.

11.9 Problems

Walking

11.1. Verify that $P(T) = \begin{cases} 0 & \text{if } S(T) \geq X, \\ X - S(T) & \text{if } S(T) < X. \end{cases}$

11.2. Prove (11.4) on p. 196.

---

[20] Subject to the boundary condition

$$F(T) = \begin{cases} 0 & \text{if } S(T) \leq X, \\ S(T) - X & \text{if } S(T) > X. \end{cases}$$

[21] Subject to the boundary condition

$$F(T) = \begin{cases} 0 & \text{if } S(T) \geq X, \\ X - S(T) & \text{if } S(T) < X. \end{cases}$$

11.3. Draw the profit-loss diagram for Helen's purchase of puts described on p. 200.

11.4. Prove the profit per share formula for writing uncovered calls on p. 202. Confirm that Fig. 11.6 is correct.

11.5. Prove the profit per share formula for writing covered calls on p. 203, and draw the corresponding profit-loss diagram.

11.6. Prove the profit per share formula for writing uncovered puts on p. 204. Confirm that Fig. 11.9 is correct.

11.7. Prove the profit per share formula for writing covered puts on p. 206. Confirm that Fig. 11.11 is correct.

11.8. Referring to the stock MOVE on p. 207, create a profit-loss diagram for a writer of a straddle for MOVE.

11.9. Prove the profit per share formula for writing straddles on p. 208. Confirm that Fig. 11.13 is correct.

11.10. Verify that if $C + Xe^{-i^{(\infty)}T} > S(0) + P$, then the position discussed on p. 210 creates an arbitrage opportunity.

11.11. BIG1 is currently trading at \$100 per share. A put on BIG1 costs \$7, the exercise price is \$105, the time until expiration is 6 months, and the continuously compounded risk-free rate of return is 4%. What should the price of a call option be to satisfy the put-call parity relationship?

11.12. Let $X = \$100$, $C = \$8$, $P = \$2$, $S(0) = \$105$, $i^{(\infty)} = 0.05$, and $T = 1$ (one year). Does the Put-Call Parity relationship hold? If not, describe an arbitrage portfolio that would generate a guaranteed positive profit. What is the amount of the profit?

11.13. Describe the arbitrage portfolio that one can create if

$$C + Xe^{-i^{(\infty)}T} > S(0) + P.$$

11.14. There is a more general put-call parity relationship for European options in the case of dividends. The condition is

$$C + PV(X) + PV(\text{dividends}) = S(0) + P.$$

Here PV stands for the present value. Assume that only one dividend payment is made, for D dollars, at expiration. Derive the appropriate relationship in this case.

11.15. The Put-Call Parity Theorem on p. 209 holds for European options, where the option can only be exercised at expiration. Are there any cases where a violation of the put-call parity relationship gives an arbitrage opportunity for American options? [Hint: Recall that American call options are never exercised early, whereas American put options may be exercised in certain cases.]

11.16. Referring to the example on p. 212 with GOUP stock, and assuming the same conditions, suppose that we have a put option on GOUP with the same exercise price ($75) and expiration time (one year) as the call option. Create a portfolio consisting of Δ^* shares of GOUP stock and B dollars of cash, discounted to the present time, which has the same payoff as a portfolio consisting of one put option.

11.17. Prove that

$$\Delta = \frac{C_U - C_D}{S_U - S_D}, \text{ and } B = \frac{C_U S_D - C_D S_U}{S_U - S_D}.$$

(See p. 213.)

11.18. Verify that the hedge ratio for a call option can never be negative. What is the value of Δ when both $S_D < X$ and $S_U < X$? What conclusions do you infer from this answer?

11.19. Verify the values for the binomial tree in Example 11.3 on p. 216.

11.20. This problem is based on Example 11.3 on p. 216. Assume that $p = 0.6$. If the stock decreases in price for the first two months, so that its price after two months is $96.04, then what is the expected price at expiration? [Hint: The possible prices at expiration are $99.92, $96.00, and $92.24. Calculate the probabilities of attaining these prices and consider the appropriate random variable.]

11.21. Prove that

$$p = \frac{S(t_0)e^{i(\infty)\delta t} - S_D(t_1)}{S_U(t_1) - S_D(t_1)},$$

for the standard binomial tree.

11.22. Consider an n-step standard binomial tree with initial stock price $S(t_0) = \$200$. Suppose that the expected value of the stock price after one month is $E(S(t_1)) = \$201$.

(a) What is the expected value of the stock price after six months?
(b) What is the expected value of the stock price after one year?
(c) What is the continuously compounded rate of return?
(d) What is the annually compounded rate of return?
(e) What is the semi-annually compounded rate of return?

Running

11.23. For the standard binomial tree, let $p = 0.5$, $0 < d \leq u$, with $u = 1/d$ (so that $ud = 1$) and an initial stock price of $S(t_0)$.

(a) What is the expected value of $S(t_1)$, $E(S(t_1))$?
(b) What is the smallest possible value of $E(S(t_1))$?

11.24. Consider the standard binomial tree for the limiting case with $d = 0$.

(a) What is the value of $E(S(t_n))$, $n \geq 1$?
(b) What are the possible values of $\lim_{n \to \infty} E(S(t_n))$?
(c) Under what conditions is $\lim_{n \to \infty} E(S(t_n)) = 0$?

11.25. Verify the call option prices for the tree on p. 222.

11.26. Calculate the call option prices for the tree discussed on p. 224.

11.27. In Problem A.4 on p. 252 you are asked to prove that for any two real numbers x and y, and n a positive integer,

$$(x + y)^n = \sum_{k=0}^{n} \binom{n}{k} x^k y^{n-k}.$$

How does this result relate to the binomial distribution?

11.28. Consider the following three-step binomial tree for the stock of company SLOW. Assume that the exercise price is $200 and that the continuously compounded rate of return is 8%. Create the corresponding tree for the call option. What can you say about the values of p?

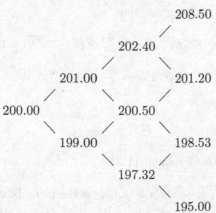

11.29. Let X be a lognormal random variable with $Y = \ln X$ having a normal distribution with mean μ and variance σ^2. Show that $E(X) = e^{\mu + \sigma^2/2}$ and

$$\sigma_X^2 = e^{2\mu + 2\sigma^2} - e^{2\mu + \sigma^2}.$$

(See p. 225.)

11.30. Show that $S(t)$ has a lognormal distribution, with $E(S(t)) = e^{i(\infty)t}S(0)$.
(See p. 225.)

11.31. What should the Black–Scholes option price be in (11.7) on p. 228, if we let $T = 0$ but do not change the other values? Verify this by taking the limit as T approaches 0 of δ (with the other values fixed) in the three cases $X < S(0)$, $X = S(0)$, and $X > S(0)$. (Note that $\Phi(\infty) = 1$, $\Phi(-\infty) = 0$.)

11.32. Show that

$$\Phi(x) = \frac{1}{\sqrt{2\pi}} \int_{-\infty}^{x} e^{-t^2/2} \, dt,$$

is defined for all x and has the following properties:

(a) $\Phi(x) > 0$.
(b) $\Phi(-x) = 1 - \Phi(x)$.
(c) $\Phi(x)$ is an increasing function.
(d) $\Phi(x)$ is concave down for $x > 0$, and concave up for $x < 0$.

11.33. Prove

$$P(0) = -S(0)\Phi(-\delta) + e^{-i(\infty)T}\Phi\left(-\delta + \sigma\sqrt{T}\right).$$

(See p. 230.)

11.34. If

$$\delta = \frac{\left(i^{(\infty)} + \frac{1}{2}\sigma^2\right)T - \ln(X/S(0))}{\sigma\sqrt{T}},$$

(see (11.8) on p. 228), then show that

$$S(0)\Phi'(\delta) - Xe^{-i(\infty)T}\Phi'(\delta - \sigma\sqrt{T}) = 0.$$

[Hint: Start with $\ln\left(\Phi'(\delta)/\Phi'(\delta - \sigma\sqrt{T})\right)$.]

11.35. Show that

$$\Phi(\delta) = \frac{\partial C(0)}{\partial S(0)}.$$

(See p. 230.) [Hint: Use Problem 11.34.]

11.36. Show that $\Phi\left(\delta - \sigma\sqrt{T}\right)$ is equal to the probability that the option is in the money at expiration, that is, $S(T) > X$. (See p. 230.)

11.37. Show that

$$\frac{\partial^2 C(0)}{\partial S(0)^2} = \frac{\Phi'(\delta)}{S(0)\sigma\sqrt{T}}.$$

(See p. 236.)

11.38. Show that

$$\frac{\partial C(0)}{\partial T} = \frac{S(0)\sigma}{2\sqrt{T}}\Phi'(\delta) + i^{(\infty)}Xe^{-i^{(\infty)}T}\Phi(\delta - \sigma\sqrt{T}).$$

(See p. 237.)

11.39. Show that

$$\frac{\partial C(0)}{\partial i^{(\infty)}} = XTe^{-i^{(\infty)}T}\Phi(\delta - \sigma\sqrt{T}).$$

(See p. 237.)

11.40. Show that

$$\frac{\partial C(0)}{\partial \sigma} = S(0)\sqrt{T}\Phi'(\delta).$$

(See p. 237.)

11.41. Show that

$$\frac{\partial C(0)}{\partial X} = -e^{-i^{(\infty)}T}\Phi(\delta - \sigma\sqrt{T}).$$

(See p. 238.)

11.42. Suppose, in Example 11.4 on p 235, that we change $S(0)$ from \$98 to \$102 but do not change the other four values. Calculate the new Black-Scholes option price. Does this agree with your intuition? Verify this by calculating the Greek Δ for this example.

11.43. Change X from \$100 to \$99 in (11.7) on p. 228 and proceed as in Problem 11.42 using the appropriate Greek.

11.44. Change $i^{(\infty)}$ from 0.05 to 0.10 in (11.7) on p. 228. Calculate the new Black–Scholes option price. Does this agree with your intuition? Verify this by calculating ρ for this example.

11.45. Change T from 0.50 to 0.25 in (11.7) on p. 228 and proceed as in Problem 11.44 using the appropriate Greek.

11.46. Change σ from 0.4 to 0.5 in (11.7) on p. 228 and proceed as in Problem 11.44 using the appropriate Greek. (Intuition may not work in this case.)

11.47. Show that

$$-\Theta + \frac{1}{2}\sigma^2 S^2(0)\Gamma + i^{(\infty)}S(0)\Delta - i^{(\infty)}C(0) = 0,$$

and that

$$-\frac{\partial C(0)}{\partial T} + \frac{1}{2}\sigma^2 S^2(0)\frac{\partial^2 C(0)}{\partial S(0)^2} + i^{(\infty)}S(0)\frac{\partial C(0)}{\partial S(0)} - i^{(\infty)}C(0) = 0.$$

(See p. 238.)

11.48. Show that the Black-Scholes formula

$$C(0) = S(0)\Phi(\delta) - Xe^{-i^{(\infty)}T}\Phi\left(\delta - \sigma\sqrt{T}\right),$$

where

$$\delta = \frac{\left(i^{(\infty)} + \frac{1}{2}\sigma^2\right)T - \ln\left(X/S(0)\right)}{\sigma\sqrt{T}},$$

where 0 is always the current time (fixed) and T is the time remaining to expiration (variable), can be written in the form

$$C(t) = S(t)\Phi(\delta^*) - Xe^{-i^{(\infty)}(T^*-t)}\Phi\left(\delta^* - \sigma\sqrt{T^*-t}\right),$$

where

$$\delta^* = \frac{\left(i^{(\infty)} + \frac{1}{2}\sigma^2\right)(T^*-t) - \ln\left(X/S(t)\right)}{\sigma\sqrt{T^*-t}},$$

where t is the current time (variable) and T^* is the expiration date (fixed).

11.49. Show, under the conditions of Problem 11.48, that the Black-Scholes partial differential equation (11.10) is

$$\frac{\partial F}{\partial t} + \frac{1}{2}\sigma^2 S^2(t)\frac{\partial^2 F}{\partial S(t)^2} + i^{(\infty)}S(t)\frac{\partial F}{\partial S(t)} - i^{(\infty)}F = 0.$$

Questions for Review

- What is a call option?
- What is a put option?
- How do the price of the stock and time to expiration affect the price of a call (put) option?
- What is a covered call?
- What is a covered put?
- What does the Put-Call Parity Theorem say?
- What does the Hedge Ratio Theorem say?
- What is a binomial tree?
- Explain arbitrage.
- What does the Black-Scholes Theorem say? What conditions are assumed?
- What are the Greeks and what do they measure?

A

Appendix: Induction, Recurrence Relations, Inequalities

A.1 Mathematical Induction

Mathematical induction is a technique used to prove a proposition that involves an integer variable. Frequently the proposition is an identity or an inequality involving an integer n such as

$$1 + 2 + 4 + \cdots + 2^{n-1} = 2^n - 1 \text{ for } n = 1, 2, \ldots,$$

or

$$n^2 > 2n + 1 \text{ for } n = 3, 4, \ldots.$$

In both cases, we have a proposition, $P(n)$, that must be proved for $n \geq n_0$.
   The technique of mathematical induction requires two steps.

1. Prove that $P(n)$ is true for the first value of n, namely, $n = n_0$. This is the **basic step**.
2. Assume that $P(k)$ is true and prove that, as a consequence of this, $P(k+1)$ is true. This is the **inductive step**.

   If both of these steps are proved, then $P(n)$ is true for $n \geq n_0$ by the following reasoning. From the basic step we know that $P(n_0)$ is true. Now, from the inductive step with $k = n_0$ we know that $P(n_0)$ being true requires that $P(n_0 + 1)$ is true. We use this and the inductive step again to show that $P(n_0 + 2)$ is true, and so on.

Example A.1. Show by induction that

$$1 + 2 + 4 + \cdots + 2^{n-1} = 2^n - 1 \text{ for } n = 1, 2, \ldots.$$

Solution. Here $n_0 = 1$ and the proposition $P(n)$ is $1 + 2 + 4 + \cdots + 2^{n-1} = 2^n - 1$.
Basic Step: We must show that $P(n_0)$, that is, $P(1)$ is true. The statement $P(1)$ is the proposition that $1 = 2^1 - 1$, which is clearly true.

Inductive Step: We must show that $P(k+1)$ is true as a consequence of $P(k)$ being true, that is, we must show that

$$1 + 2 + 4 + \cdots + 2^k = 2^{k+1} - 1$$

is a consequence of

$$1 + 2 + 4 + \cdots + 2^{k-1} = 2^k - 1.$$

Now,

$$
\begin{aligned}
1 + 2 + 4 + \cdots + 2^k &= 1 + 2 + 4 + \cdots + 2^{k-1} + 2^k \\
&= (1 + 2 + 4 + \cdots + 2^{k-1}) + 2^k \\
&= (2^k - 1) + 2^k \\
&= 2^{k+1} - 1.
\end{aligned}
$$

By mathematical induction, the equality is valid for all integers $n \geq 1$. $\triangle$

Example A.2. Show by induction that

$$n^2 > 2n + 1 \text{ for } n = 3, 4, \ldots.$$

Solution. Here $n_0 = 3$ and the proposition $P(n)$ is $n^2 > 2n + 1$.
Basic Step: We must show that $P(n_0)$, that is, $P(3)$ is true. The statement $P(3)$ is the proposition that $3^2 > 2 \times 3 + 1$, that is, $9 > 7$, which is clearly true.
Inductive Step: We must show that $P(k+1)$ is true as a consequence of $P(k)$ being true, that is, we must show that

$$(k+1)^2 > 2(k+1) + 1$$

is a consequence of

$$k^2 > 2k + 1.$$

Now,

$$
\begin{aligned}
(k+1)^2 &= k^2 + 2k + 1 \\
&> (2k+1) + 2k + 1 \\
&= 2k + (2k + 2) \\
&> 2k + 3 \\
&= 2(k+1) + 1.
\end{aligned}
$$

By mathematical induction, the inequality is valid for all integers $n \geq 3$. $\triangle$

The process of mathematical induction is frequently compared to an infinite line of dominoes that has the property that if the nth domino falls so does the next one. This is the inductive step, and no dominoes fall until one of them, the n_0^{th}, falls. This is the basic step. If the n_0^{th} domino falls, then so do all the dominos for which $n \geq n_0$.

A.2 Recurrence Relations

Definition A.1. *A* RECURRENCE RELATION *among* P_0, P_1, P_2, ..., *is an equation of the type*

$$P_{n+m} = f(P_n, P_{n+1}, \ldots, P_{n+m-1}),$$

so P_{n+m} *is determined by its previous* m *terms.*

Comments About Recurrence Relations

- Recurrence relations are also called **finite difference equations**.
- The positive integer m is the ORDER of the recurrence relation. For example, the recurrence relation

$$P_{n+2} = \frac{(n+1)P_n}{(n+3)}, \qquad n = 0, 1, 2, \ldots,$$

is a second order recurrence relation.
- There is no general way to solve recurrence relations. However, a first order recurrence relation can be solved if it can be rewritten in the form

$$Q_n = Q_{n-1}, \qquad n = 1, 2, \ldots.$$

In this form we sum both sides from 1 to N, giving

$$\sum_{n=1}^{N} Q_n = \sum_{n=1}^{N} Q_{n-1}.$$

Canceling the common terms on both sides gives

$$Q_N = Q_0,$$

so the recurrence relation $Q_n = Q_{n-1}$ has solution $Q_n = Q_0$.
(An alternative way to show this from $Q_n = Q_{n-1}$ is to realize that this implies that $Q_1 = Q_0$, that $Q_2 = Q_1$, and so on, so that

$$Q_n = Q_{n-1} = Q_{n-2} = \cdots = Q_2 = Q_1 = Q_0,$$

and so the solution is $Q_n = Q_0$.)

Example A.3. Solve the following recurrence relations ($n = 1, 2, \ldots$).

(a) $P_n = -P_{n-1}$
(b) $P_n = aP_{n-1}$, where a is constant
(c) $nP_n = P_{n-1}$
(d) $P_n = (n+1)P_{n-1}$
(e) $P_n = P_{n-1} + a$, where $a \neq 0$ is constant
(f) $P_n = P_{n-1} + ba^n$, where $a \neq 0, b \neq 0$ are constants

Solution.

(a) The recurrence relation
$$P_n = -P_{n-1}$$
can be rewritten in the form $(-1)^n P_n = (-1)^{n-1} P_{n-1}$, that is, $Q_n = Q_{n-1}$ with $Q_n = (-1)^n P_n$. This has the solution $Q_n = Q_0$, that is, $(-1)^n P_n = (-1)^0 P_0$, so $P_n = (-1)^n P_0$.

(b) The recurrence relation
$$P_n = a P_{n-1}$$
can be rewritten in the form $P_n/a^n = P_{n-1}/a^{n-1}$, that is, $Q_n = Q_{n-1}$ with $Q_n = P_n/a^n$. This has the solution $Q_n = Q_0$, that is, $P_n/a^n = P_0/a^0$, so $P_n = a^n P_0$.

(c) The recurrence relation
$$n P_n = P_{n-1}$$
can be rewritten in the form $n! P_n = (n-1)! P_{n-1}$, that is, $Q_n = Q_{n-1}$ with $Q_n = n! P_n$. This has the solution $Q_n = Q_0$, that is, $n! P_n = 0! P_0$, so $P_n = P_0/n!$.

(d) The recurrence relation
$$P_n = (n+1) P_{n-1}$$
can be rewritten in the form $P_n/(n+1)! = P_{n-1}/n!$, that is, $Q_n = Q_{n-1}$ with $Q_n = P_n/(n+1)!$. This has the solution $Q_n = Q_0$, that is, $P_n/(n+1)! = P_0/1!$, so $P_n = (n+1)! P_0$.

(e) The recurrence relation
$$P_n = P_{n-1} + a$$
can be rewritten in the form $P_n - a = P_{n-1}$ or $P_n - na = P_{n-1} - (n-1)a$, that is, $Q_n = Q_{n-1}$ with $Q_n = P_n - na$. This has the solution $Q_n = Q_0$, that is, $P_n - na = P_0 - 0a$, so $P_n = na + P_0$.

(f) The recurrence relation
$$P_n = P_{n-1} + ba^n$$
can be rewritten in the form $P_n - ba^n = P_{n-1}$, or
$$P_n - ba^n - b\sum_{i=0}^{n-1} a^i = P_{n-1} - b\sum_{i=0}^{n-1} a^i,$$
or
$$P_n - b\sum_{i=0}^{n} a^i = P_{n-1} - b\sum_{i=0}^{n-1} a^i,$$
that is, $Q_n = Q_{n-1}$ with $Q_n = P_n - b\sum_{i=0}^{n} a^i$. This has the solution $Q_n = Q_0$, that is, $P_n - b\sum_{i=0}^{n} a^i = P_0 - b$, so $P_n = b\sum_{i=0}^{n} a^i + P_0 - b = b\sum_{i=1}^{n} a^i + P_0$.

$\triangle$

A.3 Inequalities

Theorem A.1. *Arithmetic-Geometric Mean Inequality.*
If $a_1, a_2, \ldots, a_n$ are non-negative and not all zero, then

$$(a_1 a_2 \cdots a_n)^{1/n} \leq \frac{a_1 + a_2 + \cdots + a_n}{n},$$

with equality if and only if $a_1 = a_2 = \cdots = a_n$.

Proof. The proof of the Arithmetic-Geometric Mean Inequality proceeds in two stages. First, we prove it by induction for $n = 2^m$, namely,

$$(a_1 a_2 \cdots a_{2^m})^{1/2^m} \leq \frac{a_1 + a_2 + \cdots + a_{2^m}}{2^m}$$

and then we fill in the missing n's.

With $m = 1$ we want to show that

$$(a_1 a_2)^{1/2} \leq \frac{a_1 + a_2}{2}.$$

However, from $(a_1 - a_2)^2 \geq 0$, we have $a_1^2 + a_2^2 \geq 2 a_1 a_2$, which is the required inequality. Equality occurs if and only if $a_1 = a_2$.

We assume that this is valid for m and show that it is true for $m + 1$, that is, we want to show that

$$(a_1 a_2 \cdots a_{2^{m+1}})^{1/2^{m+1}} \leq \frac{a_1 + a_2 + \cdots + a_{2^{m+1}}}{2^{m+1}},$$

with equality if and only if $a_1 = a_2 = \cdots = a_{2^{m+1}}$.

Now,

$$(a_1 a_2 \cdots a_{2^{m+1}})^{1/2^{m+1}} = \left[(a_1 a_2 \cdots a_{2^m})^{1/2^m} (a_{2^m+1} a_{2^m+2} \cdots a_{2^{m+1}})^{1/2^m} \right]^{1/2}.$$

But,

$$\left[(a_1 a_2 \cdots a_{2^m})^{1/2^m} (a_{2^m+1} a_{2^m+2} \cdots a_{2^{m+1}})^{1/2^m} \right]^{1/2}$$
$$\leq \left[\left(\tfrac{a_1 + a_2 + \cdots + a_{2^m}}{2^m} \right) \left(\tfrac{a_{2^m+1} + a_{2^m+2} + \cdots + a_{2^{m+1}}}{2^m} \right) \right]^{1/2},$$

with equality if and only if $a_1 = a_2 = \cdots = a_{2^m}$ and $a_{2^m+1} = a_{2^m+2} = \cdots = a_{2^{m+1}}$. Now,

$$\left[\left(\tfrac{a_1 + a_2 + \cdots + a_{2^m}}{2^m} \right) \left(\tfrac{a_{2^m+1} + a_{2^m+2} + \cdots + a_{2^{m+1}}}{2^m} \right) \right]^{1/2}$$
$$\leq \tfrac{1}{2} \left[\left(\tfrac{a_1 + a_2 + \cdots + a_{2^m}}{2^m} \right) + \left(\tfrac{a_{2^m+1} + a_{2^m+2} + \cdots + a_{2^{m+1}}}{2^m} \right) \right],$$

with equality if and only if $a_1 + a_2 + \cdots + a_{2^m} = a_{2^m+1} + a_{2^m+2} + \cdots + a_{2^{m+1}}$. This is the desired inequality, with equality if and only if $a_1 = a_2 = \cdots = a_{2^{m+1}}$.

Now, consider the case when $n \neq 2^m$. Then choose $a_{n+1} = a_{n+2} = \cdots = a_{2^m} = A$, where

$$A = \frac{a_1 + a_2 + \cdots + a_n}{n}.$$

Thus,

$$\left(a_1 a_2 \cdots a_n A^{2^m - n}\right)^{1/2^m} \leq \frac{a_1 + a_2 + \cdots + a_n + (2^m - n) A}{2^m} = A,$$

or

$$a_1 a_2 \cdots a_n A^{2^m - n} \leq A^{2^m},$$

that is,

$$a_1 a_2 \cdots a_n \leq A^n.$$

Equality occurs if and only if $a_1 = a_2 = \cdots = a_n$.

This completes the proof. $\qquad\square$

Theorem A.2.

$$\left(1 + \frac{r}{n-1}\right)^{n-1} < \left(1 + \frac{r}{n}\right)^n.$$

Proof. To prove this we use the Arithmetic-Geometric Mean Inequality with $a_1 = 1$, $a_2 = a_3 = \cdots = a_n = 1 + r/(n-1)$, which gives

$$\left[1\left(1 + \frac{r}{n-1}\right)^{n-1}\right]^{1/n} \leq \frac{1 + (n-1)\left(1 + \frac{r}{n-1}\right)}{n},$$

or

$$\left[\left(1 + \frac{r}{n-1}\right)^{n-1}\right]^{1/n} \leq \frac{n+r}{n} = 1 + \frac{r}{n}.$$

$\qquad\square$

Theorem A.3. *Cauchy-Schwarz Inequality.*

$$\left(\sum_{i=1}^{n} a_i b_i\right)^2 \leq \left(\sum_{i=1}^{n} a_i^2\right)\left(\sum_{i=1}^{n} b_i^2\right),$$

with equality if and only if either $a_i = \lambda b_i$ for some constant λ or $b_i = 0$ for all i.

Proof. The Cauchy-Schwarz inequality is obviously true if $b_1 = b_2 = \cdots = b_n = 0$, so we concentrate on the case when not all b_i are zero, in which case $\sum_{j=1}^{n} b_j^2 \neq 0$. Consider, with

$$\lambda = \frac{\sum_{j=1}^{n} a_j b_j}{\sum_{j=1}^{n} b_j^2},$$

the quantity

$$0 \le \sum_{i=1}^{n}(a_i - \lambda b_i)^2 = \sum_{i=1}^{n}a_i^2 - 2\lambda\sum_{i=1}^{n}a_ib_i + \lambda^2\sum_{i=1}^{n}b_i^2.$$

Clearly the equality holds if and only if $a_i = \lambda b_i$ for all i. Otherwise we have

$$0 < \sum_{i=1}^{n}a_i^2 - 2\frac{\sum_{j=1}^{n}a_jb_j}{\sum_{j=1}^{n}b_j^2}\sum_{i=1}^{n}a_ib_i + \left(\frac{\sum_{j=1}^{n}a_jb_j}{\sum_{j=1}^{n}b_j^2}\right)^2\sum_{i=1}^{n}b_i^2,$$

which can be written as

$$0 < \sum_{i=1}^{n}a_i^2 - \frac{\left(\sum_{j=1}^{n}a_jb_j\right)^2}{\sum_{j=1}^{n}b_j^2}.$$

This is the Cauchy-Schwarz inequality. $\qquad\square$

Theorem A.4. *Hölder's Inequality.*
For $p > 1$ and q such that $1/p + 1/q = 1$,

$$\left|\sum_{i=1}^{n}u_iv_i\right| \le \left(\sum_{i=1}^{n}|u_i|^p\right)^{1/p}\left(\sum_{i=1}^{n}|v_i|^q\right)^{1/q}.$$

Proof. We assume, without loss of generality, that at least one value of u_i and at least one value of v_i is non-zero. We may also assume that $u_i \ge 0$ and that $v_i \ge 0$ for all i. (Why?) Let $p > 0$ and define q so that $1/p+1/q = 1$. For fixed $y > 0$, consider the function $f(x) = xy - x^p/p$. We have $f'(x) = y - x^{p-1}$, so $f'(x) = 0$ if and only if $x = y^{1/(p-1)}$. For this value of x,

$$f(x) = y^{1/(p-1)+1} - \frac{y^{p/(p-1)}}{p} = y^{p/(p-1)}\left(1 - \frac{1}{p}\right) = \frac{y^q}{q}.$$

Thus, for all $x \ge 0$, $xy \le x^p/p + y^q/q$. Now, let

$$x_i = \frac{u_i}{\left(\sum_{i=1}^{n}|u_i|^p\right)^{1/p}}$$

and

$$y_i = \frac{v_i}{\left(\sum_{i=1}^{n}|v_i|^q\right)^{1/q}},$$

so that

$$u_iv_i \le \left(\frac{1}{p}\frac{|u_i|^p}{\sum_{i=1}^{n}|u_i|^p} + \frac{1}{q}\frac{|v_i|^q}{\sum_{i=1}^{n}|v_i|^q}\right)\left(\sum_{i=1}^{n}|u_i|^p\right)^{1/p}\left(\sum_{i=1}^{n}|v_i|^q\right)^{1/q}.$$

Summing from $i = 1, 2, \ldots, n$ gives

$$\sum_{i=1}^{n} u_i v_i \leq \left(\frac{1}{p} + \frac{1}{q}\right) \left(\sum_{i=1}^{n} |u_i|^p\right)^{1/p} \left(\sum_{i=1}^{n} |v_i|^q\right)^{1/q}$$

$$= \left(\sum_{i=1}^{n} |u_i|^p\right)^{1/p} \left(\sum_{i=1}^{n} |v_i|^q\right)^{1/q}.$$

Hölder's Inequality follows.   □

Note that the main statement in the Cauchy-Schwarz Inequality is a special case of Hölder's Inequality with $p = q = 2$.

A.4 Problems

A.1. Prove that

$$\sum_{k=1}^{n} k = \frac{1}{2} n (n + 1)$$

for $n = 1, 2, \ldots$ by induction.

A.2. Prove that

$$\sum_{k=1}^{n} k^2 = \frac{1}{6} n (n + 1) (2n + 1)$$

for $n = 1, 2, \ldots$ by induction.

A.3. Prove that

$$\sum_{k=1}^{n} k^3 = \frac{1}{4} n^2 (n + 1)^2$$

for $n = 1, 2, \ldots$ by induction.

A.4. Consider[1]

$$\binom{n}{k} = \frac{n!}{k! (n - k)!}.$$

(a) Prove that

$$\binom{n}{k-1} + \binom{n}{k} = \binom{n+1}{k}.$$

(b) Use part (a) and induction to prove the binomial expansion

$$(a + b)^n = \sum_{k=0}^{n} \binom{n}{k} a^k b^{n-k}.$$

---

[1] Remember that $0! = 1$ and that $n! = n (n - 1) (n - 2) \cdots 3(2)(1)$ for $n > 0$.

A.5. Use induction to prove the triangle inequality

$$\left| \sum_{k=1}^{n} x_k \right| \le \sum_{k=1}^{n} |x_k| .$$

A.6. Prove that the number of subsets[2] of a set with n elements is 2^n.

A.7. Show that $f(n) = n^2 + n + 41$ is prime for $n = 0$, $n = 1$, $n = 2$, $n = 4$, and $n = 5$. Try this for a few more values of n, say up to $n = 10$. Do you think this means that $f(n)$ is prime for all n? If so, prove it. If not, supply a counterexample, that is, find an n for which $f(n)$ is the product of two positive integers, neither of which is 1.

---

[2] Remember that the empty set is a subset of every set.

B

Appendix: Statistics

A basic knowledge of probability and statistics is necessary for understanding much of the material in Chaps. 10 and 11. Two useful references are [13] and [20].

B.1 Set Theory

Consider an experiment (such as the tossing of a coin) that may result in a fixed (possibly infinite) number of outcomes. The set of all outcomes is called the sample space of the experiment. An event is a subset of the sample space. Events are said to be mutually exclusive (or disjoint) if they have no elements in common.

For any two events A and B we have the following definitions:

1. The union of A and B, $A \cup B$, is the set of all elements that belong to A or to B (or both A and B).
2. The intersection of A and B, $A \cap B$, is the set of all elements that belong to both A and B.
3. The complement of A, A^c, is the set of all elements that do not belong to A.
4. The empty set, $\emptyset$, is the set consisting of no elements.
5. A is a subset of B, written $A \subset B$, if every element of A is also an element of B.
6. $A = B$ if $A \subset B$ and $B \subset A$.

We extend 1 and 2 as follows:

1' If $A_1, A_2, \ldots, A_n$ are events, then $A_1 \cup A_2 \cup \cdots \cup A_n$ is the event consisting of all elements that are in at least one of the events $A_1, A_2, \ldots, A_n$.
2' If $A_1, A_2, \ldots, A_n$ are events, then $A_1 \cap A_2 \cap \cdots \cap A_n$ is the event consisting of all elements that belong to each of the sets $A_1, A_2, \ldots, A_n$.

1' and 2' extend in the obvious way for a countably infinite number of events $A_1, A_2, \ldots$.

B.2 Probability

A probability measure is a real-valued function $P(\cdot)$ defined on a set of events such that

1. $0 \leq P(A) \leq 1$ for any event A.
2. $P(A_1 \cup A_2 \cup \cdots) = P(A_1) + P(A_2) + \cdots$ if $A_1, A_2, \ldots$ are mutually exclusive events.
3. $P(S) = 1$, where S is the sample space.

Note that from 2 we have $P(\emptyset) = 2P(\emptyset)$, implying that $P(\emptyset) = 0$.

Because $A \cup A^c = S$ and A and A^c are mutually exclusive, we have $P(A) + P(A^c) = P(S)$, so $P(A^c) = 1 - P(A)$ for any event A.

For two events A and B, with $P(B) > 0$, we define the conditional probability of A given B by $P(A|B) = P(A \cap B)/P(B)$.

Consider the following example: Suppose that $P(A \cap B) = 0.10$, $P(A) = 0.40$, and $P(B) = 0.30$. We have $P(A|B) = P(A \cap B)/P(B) = 0.10/0.30 = 0.33$ and $P(B|A) = P(A \cap B)/P(A) = 0.25$.

We say that A and B are independent events if $P(A \cap B) = P(A)P(B)$. If A and B are independent and $P(B) > 0$, then we have $P(A|B) = P(A \cap B)/P(B) = (P(A)P(B))/P(B) = P(A)$, so A and B are independent if $P(A|B) = P(A)$.

We extend the definition of independence as follows: Events $A_1, A_2, \ldots, A_n$ are said to be independent if for $l = 2, \ldots, n$, and distinct indices $i_1, i_2, \ldots, i_l \subset \{1, \ldots, n\}$, $P(A_{i_1} \cap A_{i_2} \cap \cdots \cap A_{i_l}) = P(A_{i_1})P(A_{i_2}) \cdots P(A_{i_l})$.

Consider the following example: Let $P(A_1 \cap A_2 \cap A_3) = 0.125$, $P(A_1 \cap A_2) = P(A_1 \cap A_3) = P(A_2 \cap A_3) = 0.25$, and $P(A_1) = P(A_2) = P(A_3) = 0.50$. It is easy to demonstrate that A_1, A_2, and A_3 are independent events.

The definition of independence extends in the obvious way for a countably infinite number of events $A_1, A_2, \ldots$.

B.3 Random Variables

A random variable is a real-valued function defined on the sample space. The probability distribution of the random variable is determined by its distribution function. The DISTRIBUTION FUNCTION OF A RANDOM VARIABLE X is given by

$$F(x) = P(X \leq x),$$

defined for any real number x.

Example B.1. Consider the stock of LMN Corp, where X is the hourly dollar change in the stock price. Assume that the probability distribution of X is as follows: $P(X = -0.50) = 0.15$, $P(X = -0.25) = 0.20$, $P(X = 0) = 0.20$, $P(X = 0.25) = 0.20$, and $P(X = 0.50) = 0.25$. Find the probabilities of an hourly increase and an hourly decrease in the stock price.

Solution. The event $(X > 0)$ is the event that there is an hourly increase in the stock price. Note that $P(X > 0) = P(X = 0.25) + P(X = 0.50) = 0.20 + 0.25 = 0.45$. This is the probability of an hourly increase in the stock price. Also, $P(X < 0) = P(X = -0.50) + P(X = -0.25) = 0.15 + 0.20 = 0.35$. This is the probability of an hourly decrease in the stock price. Thus, the stock price is more likely to increase than to decrease. $\triangle$

B.3.1 Discrete Random Variables

We say that a random variable is a DISCRETE RANDOM VARIABLE if the set of all possible values of the random variable is countable. The random variable X in Example B.1 is a discrete random variable. For any discrete random variable X the probability function of X, $f(x)$, is defined for all real numbers x by $f(x) = P(X = x)$. Note that $f(x)$ is a probability function if and only if $f(x) \geq 0$ for all x and

$$\sum_{all \ x} f(x) = 1.$$

Example B.2. Find the probability function of the random variable X in Example B.1.

Solution. The probability function is given by $f(-0.50) = 0.15$, $f(-0.25) = 0.20$, $f(0) = 0.20$, $f(0.25) = 0.20$, $f(0.50) = 0.25$, and $f(x) = 0.00$ otherwise. $\triangle$

We may be interested in the average hourly price change of LMN Corp. This leads to the following definition: For a discrete random variable X with possible values $x_1, x_2, \ldots$, we define the mean or expected value of X by

$$\mu = E(X) = \sum_{all \ x} xf(x),$$

if the sum is defined. Note that μ is a weighted average of the possible values of X, where the weights $f(x)$ sum to one.

For LMN Corp. we have

$$\mu = E(X)$$
$$= -0.50 \times 0.15 - 0.25 \times 0.20 + 0.0 \times 0.20 + 0.25 \times 0.20 + 0.50 \times 0.25$$
$$= 0.05,$$

which is 5 cents. Note that this is not the unweighted or equally weighted average of the possible values of X, which is $-0.50 - 0.25 + 0.0 + 0.25 + 0.50 - 0.0$.

Example B.3. Let Y be the random variable that gives the hourly dollar price change in the stock of company OPQ. This random variable has the following probability distribution.

y	$f(y)$
-0.5	0.20
0.0	0.50
0.5	0.30

Find the expected value of Y.

Solution. The expected value is $E(Y) = -0.50 \times 0.20 + 0 \times 0.50 + 0.50 \times 0.30 = 0.05$, which is 5 cents. $\triangle$

Notice that the mean hourly dollar price change is the same for companies LMN and OPQ. However, the probability distributions of the two random variables are quite different. For example, $P(Y = -0.50) = 0.20 > P(X = -0.50) = 0.15$, and $P(Y = 0.50) = 0.30 > P(X = 0.50) = 0.25$. Thus, it is more probable that Y takes on the extreme values -0.50 and 0.50.

To account for the variability in the possible values of a random variable we have the following definitions:

If X is a discrete random variable with finite mean $E(X)$, then we define the variance of X by

$$\sigma^2 = E\left((X - \mu)^2\right) = \sum_{all\ x} (x - \mu)^2 f(x).$$

Note that σ^2 may be infinite and that $\sigma^2 = 0$ if and only if $f(\mu) = 1$. The standard deviation of X is $\sigma = \sqrt{\sigma^2}$. Note that

$$\sigma^2 = \sum_{all\ x} (x - \mu)^2 f(x)$$
$$= \sum_{all\ x} \left(x^2 - 2\mu x + \mu^2\right) f(x)$$
$$= \sum_{all\ x} x^2 f(x) - 2\mu \sum_{all\ x} x f(x) + \mu^2 \sum_{all\ x} f(x)$$
$$= \sum_{all\ x} x^2 f(x) - 2\mu^2 + \mu^2$$
$$= \sum_{all\ x} x^2 f(x) - \mu^2.$$

For LMN Corp. we have

$$\sigma^2 = (-0.50 - 0.05)^2 \times 0.15 + (-0.25 - 0.05)^2 \times 0.20$$
$$+ (0.00 - 0.05)^2 \times 0.20 + (0.25 - 0.05)^2 \times 0.20 + (0.50 - 0.05)^2 \times 0.25$$
$$= 0.1225.$$

We may also compute σ^2 as follows:

$$\sigma^2 = \sum_{all\ x} x^2 f(x) - \mu^2$$
$$= (-0.50)^2 \times 0.15 + (-0.25)^2 \times 0.20 + (0)^2 \times 0.20 + (0.25)^2 \times 0.20$$
$$+ (0.50)^2 \times 0.25 - (0.05)^2$$
$$= 0.1225.$$

It is left as an exercise to show that the variance of the random variable Y is also equal to 0.1225. This shows that two discrete random variables may have the same means and variances, but different probability distributions.

B.3.2 Independence of Random Variables

If $X_1, X_2, \ldots, X_n$ are random variables defined on the same sample space, then we say that they are INDEPENDENT if

$$P(X_1 \le x_1, X_2 \le x_2, \ldots, X_n \le x_n)$$
$$= P(X_1 \le x_1)P(X_2 \le x_2) \cdots P(X_n \le x_n)$$

for any real numbers $x_1, x_2, \ldots, x_n$. An infinite collection of random variables is said to be independent if every finite collection of the random variables is independent.

It can be shown that if $X_1, X_2, \ldots, X_n$ are independent random variables and $A_1, A_2, \ldots, A_n$ are subsets of real numbers such that the events $(X_1 \subset A_1), (X_2 \subset A_2), \ldots, (X_n \subset A_n)$ are defined, then

$$P(X_1 \subset A_1, X_2 \subset A_2, \ldots, X_n \subset A_n)$$
$$= P(X_1 \subset A_1)P(X_2 \subset A_2) \cdots P(X_n \subset A_n).$$

Furthermore, if $f_1, f_2, \ldots, f_n$ are functions such that $f_1(X_1), f_2(X_2), \ldots, f_n(X_n)$ are random variables, then $f_1(X_1), f_2(X_2), \ldots, f_n(X_n)$ are independent random variables.

Also if $X_1, X_2, \ldots, X_n$ are independent random variables and $E(|X_i|^n)$ is finite for $i = 1, 2, \ldots, n$, then

$$E(X_1 X_2 \cdots X_n) = E(X_1)E(X_2) \cdots E(X_n).$$

Now, consider the following experiment as exemplified by the repeated tossing of a (possibly biased) coin:

1. The number, n, of trials is fixed in advance.
2. Each trial results in either a success or a failure.
3. The probability of success is p, with $0 \le p \le 1$, on each trial.
4. The trials are independent.

We define the random variables $X_1, X_2, \ldots, X_n$ by $X_i = 1$ if the ith trial is a success, and $X_i = 0$ if the ith trial is a failure. We see that $X_1, X_2, \ldots, X_n$ are independent.

Let $X = \sum_{i=1}^{n} X_i$. Then X is the number of successes in the n trials. A random variable satisfying conditions 1 through 4 is a BINOMIAL RANDOM VARIABLE. Note that the probability distribution of X is completely determined by the parameters n and p. We now determine the probability distribution and mean and variance of X.

Theorem B.1. *If X is a binomial random variable with parameters n and p, then we have*

(a) $f(x) = P(X = x) = \binom{n}{x} p^x (1 - p)^{n-x}$, $x = 0, \ldots, n$.
(b) $\mu = E(X) = np$.
(c) $\sigma^2 = E\left((X - \mu)^2\right) = np(1 - p)$.

Proof.

(a) We prove part (a) using induction on k, the number of trials. Now, (a) is true for $k = 1$ because in this case $f(0) = P(X = 0) = 1 - p$ and $f(1) = P(X = 1) = p$. Now assume that (a) holds for $k = 1, \ldots, n - 1$. We need to show that (a) holds for $k = n$. In this case, $f(x) = P(X = x) = P(x$ successes in the first $n - 1$ trials and a failure on the nth trial$) + P(x - 1$ successes in the first $n - 1$ trials and a success on the nth trial$)$ for $x = 0, \ldots, n$. (Note that the second probability is zero if $x = 0$.) Using independence,

$$
\begin{aligned}
f(x) &= P(X = x) \\
&= \binom{n-1}{x} p^x (1 - p)^{(n-1)-x} (1 - p) + \binom{n-1}{x-1} p^{x-1} (1 - p)^{(n-1)-(x-1)} p \\
&= \left[\binom{n-1}{x} + \binom{n-1}{x-1} \right] p^x (1 - p)^{n-x}.
\end{aligned}
$$

We know, from Problem A.4 on p. 252, that

$$
\binom{n-1}{x} + \binom{n-1}{x-1} = \binom{n}{x},
$$

so

$$
f(x) = P(X = x) = \binom{n}{x} p^x (1 - p)^{n-x}.
$$

This proves part (a).

(b) We prove part (b) using algebraic manipulation. The mean of X is given by

$$\mu = E(X) = \sum_{all\ x} xf(x)$$

$$= \sum_{x=0}^{n} x \binom{n}{x} p^x (1-p)^{n-x}$$

$$= \sum_{x=0}^{n} x \frac{n!}{x!\,(n-x)!} p^x (1-p)^{n-x}$$

$$= \sum_{x=1}^{n} x \frac{n!}{x!\,(n-x)!} p^x (1-p)^{n-x}$$

$$= np \sum_{x=1}^{n} \frac{(n-1)!}{(x-1)!\,(n-x)!} p^{x-1} (1-p)^{n-x}.$$

Let $y = x - 1$. Then

$$E(X) = np \sum_{y=0}^{n-1} \binom{n-1}{y} p^y (1-p)^{(n-1)-y}$$

$$= np.$$

Here we used the fact that $\binom{n-1}{y} p^y (1-p)^{(n-1)-y}$ is the probability that a binomial random variable with parameters $n-1$ and p is equal to the value y, $y = 0, 1, \ldots, n-1$.

(c) The proof of part (c) is left to you (see Problem B.2).   □

Consider the following three binomial distributions. In each case the number of trials, n, is equal to four, but the probability p goes from 0.1, to 0.5, to 0.9.

x	$f(x)$ $p=0.1$	$p=0.5$	$p=0.9$
0	0.6561	0.0625	0.0001
1	0.2916	-0.2500	0.0036
2	0.0486	0.3750	0.0486
3	0.0036	0.2500	0.2916
4	0.0001	0.0625	0.6561

We note that the graphs for $p = 0.1$ (Fig. B.1) and $p = 0.9$ (Fig. B.2) are mirror images and that the graph for $p = 0.5$ (Fig. B.3) is symmetric about its mean $\mu = np = 4 \times 0.50 = 2.0$. Can you see this from the table?

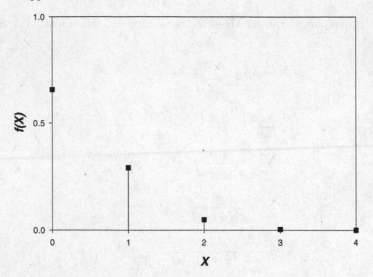

Fig. B.1. Binomial distribution with $n = 4$ and $p = 0.1$

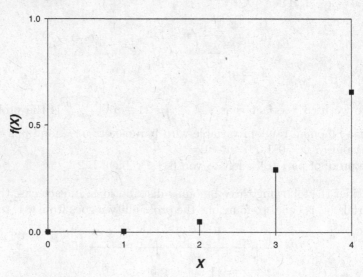

Fig. B.2. Binomial distribution with $n = 4$ and $p = 0.9$

B.3.3 Duration and Random Variables

In Chap. 8 we introduce the concept of duration as a time-weighted average of the future values of cash flows. Using the notation of that chapter, let $C(k)$ be the cash flow at period k, paid m times a year, where $0 \le k \le n$, and let y be the interest rate per period. We assume that $C(k) \ge 0$ for all k.

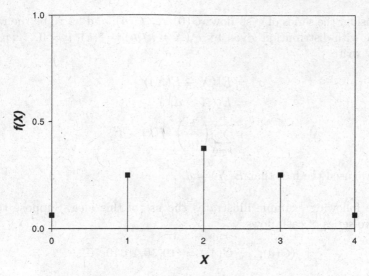

Fig. B.3. Binomial distribution with $n = 4$ and $p = 0.5$

Then we define duration as

$$d = \frac{\sum_{k=0}^{n} \frac{k}{m} C(k) (1+y)^{-k}}{\sum_{k=0}^{n} C(k) (1+y)^{-k}}.$$

We now define a random variable X with expectation $E(X) = d$.

Let $f(k)$, $0 \le k \le n$, be defined by

$$f(k) = \frac{C(k) (1+y)^{-k}}{\sum_{k=0}^{n} C(k) (1+y)^{-k}}.$$

Then $0 \le f(k) \le 1$, $0 \le k \le n$, and $\sum_{k=0}^{n} f(k) = 1$, so the $f(k)$'s may be thought of as probabilities. Now, let X be a random variable satisfying $P(X = k/m) = f(k)$, $k = 0, \ldots, n$. By construction $E(X) = d$.

Consider two different series of cash flows:

$$C(0), \ldots, C(n) \text{ and } D(0), \ldots, D(n).$$

Suppose that $C(k) \ge 0$ and $D(k) \ge 0$, $k = 0, \ldots, n$, that $\sum_{k=0}^{n} C(k) = \sum_{k=0}^{n} D(k)$, that the cash flows are paid m times a year, that the interest rate per period y is the same for both sets of cash flows, and that the durations of the two sets of cash flows are the same.

What makes an investor favor one series of cash flows over the other? Viewing the cash flows in terms of random variables, we derive a measure of the "variability" of the cash flows, which may help in deciding which series of cash flows is preferable.

Consider the series of cash flows $C(0), \ldots, C(n)$, and let X be the random variable with distribution given by $P(X = k/m) = f(k)$, $k = 0, \ldots, n$. Then we may write

$$
\begin{aligned}
\sigma^2 &= E((X - E(X))^2) \\
&= E((X - d)^2) \\
&= \sum_{k=0}^{n} \left(\frac{k}{m} \right)^2 f(k) - d^2,
\end{aligned}
$$

where we used the fact that $E(X) = d$.

The following example illustrates the use of this idea. Suppose that we have two series of cash flows:

$$
(C(0), \ldots, C(4)) = (10, 10, 10, 10, 10),
$$

and

$$
(D(0), \ldots, D(4)) = (23.7809, 0, 0, 0, 26.2191).
$$

Also assume that $y = 0.05$ and $m = 2$ in both cases. Note that $\sum_{k=0}^{4} C(k) = \sum_{k=0}^{4} D(k) = 50$. Let X_1 and X_2 be the random variables associated with the two series of cash flows. For the first series of cash flows,

$$
\begin{aligned}
f(k) &= P(X_1 = k/m) \\
&= P(X_1 = k/2) \\
&= \frac{(1.05)^{-k}}{\sum_{k=0}^{4} (1.05)^{-k}}.
\end{aligned}
$$

Then $f(0) = 0.2200$, $f(1) = 0.2095$, $f(2) = 0.1995$, $f(3) = 0.1900$, $f(4) = 0.1810$, and $E(X_1)$ is

$$
d_1 = \frac{\sum_{k=0}^{n} \frac{k}{m} C(k)(1+y)^{-k}}{\sum_{k=0}^{n} C(k)(1+y)^{-k}} = 0.9513.
$$

For the second series of cash flows we have

$$
g(k) = P(X_2 = (k/2)) = \begin{cases} 0.5244, & \text{if } k = 0, \\ 0, & \text{if } k = 1, 2, 3, \\ 0.4756 & \text{if } k = 4, \end{cases}
$$

and $d_2 = 0.9513$.

Thus, both series of cash flows yield the same total cash flow, 50, and have the same duration, 0.9513. Now, we compare the variances of X_1 and X_2. We have $\sigma_{X_1}^2 = \sum_{k=0}^{4} (k/2)^2 f(k) - d_1^2 = 0.4985$. A similar calculation gives $\sigma_{X_2}^2 = 0.9975$.

Thus, the variance of X_2 is approximately twice that of X_1. This reflects the fact that there are only two positive cash flows (23.7809 and 26.2191) for the second series of cash flows, one at either end, while the first series is a constant series of cash flows (10 each period). An investor who wants a constant series of cash flows may prefer the first series, whereas an investor who wants a large cash flow immediately and is willing to delay the remaining payments may prefer the second series.

B.3.4 Continuous Random Variables

Random variables measuring quantities such as height in centimeters, weight in pounds, and length in meters are not discrete random variables because the number of possible values is not countable. In these cases we have the following definition:

A random variable X is a CONTINUOUS RANDOM VARIABLE if its distribution function, $F(x)$, is differentiable for all x. Because continuous random variables can take an uncountable number of values, probability calculations cannot be done by summing probabilities. Instead, these variables are characterized by their density functions, and probabilities are calculated using integration. The density function of a continuous random variable is defined for all real numbers x by

$$f(x) = \frac{d}{dx} F(x).$$

Note that $f(x)$ is a density function if and only if $f(x) \geq 0$ for all x and

$$\int_{-\infty}^{\infty} f(x)\, dx = 1.$$

It follows that[1]

$$F(x) = P(X \leq x) = \int_{-\infty}^{x} f(t)\, dt.$$

Consider the following example: Let X have a uniform distribution on the interval $[0, 1]$. That is, $F(x) = P(X \leq x) = 0$ if $x < 0$, $F(x) = x$ if $0 \leq x \leq 1$, and $F(x) = 1$ if $x > 1$. It is easy to see that in this case we may set $f(x) = 0$ if $x \leq 0$, $f(x) = 1$ if $0 < x < 1$, and $f(x) = 0$ if $x \geq 1$.[2]

We define the mean of a continuous random variable X with density function $f(x)$ by

$$\mu = E(X) = \int_{-\infty}^{\infty} x f(x)\, dx,$$

if the integral exists. If $E(X)$ exists, then we define the variance of X by

$$\sigma^2 = E((X - \mu)^2) = \int_{-\infty}^{\infty} (x - \mu)^2 f(x)\, dx.$$

---

[1] Because $\int_x^x f(t)\, dt = 0$, we have $P(X < x) = P(X \leq x)$.
[2] It does not matter how $f(x)$ is defined at $x = 0$ or $x = 1$.

The standard deviation is $\sigma = \sqrt{\sigma^2}$. The variance σ^2 may be infinite, but in most practical cases $0 < \sigma^2 < \infty$.

The most important continuous random variable is the NORMAL RANDOM VARIABLE, which is characterized by its mean μ and variance σ^2. It plays a basic role in modeling the rates of returns on investments. It is defined as follows:

A random variable X has a normal distribution with parameters μ and σ^2 if

$$F(x) = P(X \leq x) = \frac{1}{\sqrt{2\pi\sigma^2}} \int_{-\infty}^{x} e^{-(t-\mu)^2/(2\sigma^2)} \, dt,$$

where $-\infty < \mu < \infty$ and $0 < \sigma^2 < \infty$. Thus, the density function of a normal random variable with parameters μ and σ^2 is given by

$$f(x) = \frac{1}{\sqrt{2\pi\sigma^2}} e^{-(x-\mu)^2/(2\sigma^2)},$$

where $-\infty < x < \infty$.

Theorem B.2. *The following properties hold for a normal random variable:*

(a) $f(x)$ is a density function.
(b) The expected value of X is $E(X) = \mu$.
(c) The variance of X is $\sigma^2 = E\left((X - \mu)^2\right)$.

Proof.

(a) The proof of part (a) requires multivariate calculus and is omitted.
(b) Using the substitution $w = x - \mu$, we have

$$E\left(X\right)$$
$$= \frac{1}{\sqrt{2\pi\sigma^2}} \int_{-\infty}^{\infty} x e^{-(x-\mu)^2/(2\sigma^2)} \, dx$$
$$= \frac{1}{\sqrt{2\pi\sigma^2}} \int_{-\infty}^{\infty} (x - \mu + \mu) e^{-(x-\mu)^2/(2\sigma^2)} \, dx$$
$$= \frac{1}{\sqrt{2\pi\sigma^2}} \left(\int_{-\infty}^{\infty} (x - \mu) e^{-(x-\mu)^2/(2\sigma^2)} \, dx + \int_{-\infty}^{\infty} \mu e^{-(x-\mu)^2/(2\sigma^2)} \, dx \right)$$
$$= \frac{1}{\sqrt{2\pi\sigma^2}} \int_{-\infty}^{\infty} w e^{-w^2/(2\sigma^2)} \, dw + \frac{\mu}{\sqrt{2\pi\sigma^2}} \int_{-\infty}^{\infty} e^{-(x-\mu)^2/(2\sigma^2)} \, dx.$$

Both integrals exist, and by symmetry, the value of the first integral is 0. Because $f(x)$ is a density function the value of the second integral is μ.
(c) The proof of part (c) is left as an exercise. $\qquad\qquad\square$

Figure B.4 shows the density functions of normal random variables with common mean $\mu = 100$ and variances $\sigma^2 = 0.5$, 1.0, and 2.0. Note that the curves are all symmetric across the mean $\mu = 100$ and that they become flatter as σ^2 becomes larger.

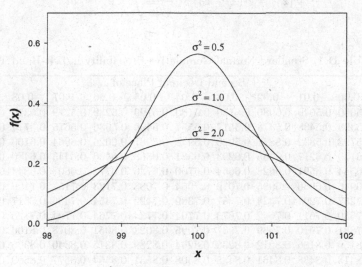

Fig. B.4. Density functions with mean $\mu = 100$ and variances $\sigma^2 = 0.5$, 1.0, and 2.0

One interesting and useful property of the normal distribution is the fact that it is invariant under linear transformations.

Theorem B.3. *If X has a normal distribution with mean μ and variance σ^2, and if a and b are any two real numbers, then the random variable $Y = aX + b$ has a normal distribution with mean $a\mu + b$ and variance $a^2\sigma^2$.*

Proof. In the following we assume, without loss of generality, that $a > 0$. Let y be any real number. Then

$$P(Y \leq y) = P(aX + b \leq y)$$
$$= P(X \leq (y - b)/a)$$
$$= \frac{1}{\sqrt{2\pi\sigma^2}} \int_{-\infty}^{(y-b)/a} e^{-(x-\mu)^2/(2\sigma^2)} \, dx.$$

Using the substitution $t = ax + b$ we have

$$P(Y \leq y) = \frac{1}{\sqrt{2\pi a^2\sigma^2}} \int_{-\infty}^{y} e^{-(t-(a\mu+b))^2/(2a^2\sigma^2)} \, dt,$$

which is the distribution function of a normal random variable with mean $a\mu + b$ and variance $a^2\sigma^2$ evaluated at y. This completes the proof of the theorem. □

Table B.1 gives values of the standard normal distribution function $F(z)$, where Z is the standard normal random variable—the normal random variable with mean $\mu = 0$ and variance $\sigma^2 = 1$. The function $F(z_0)$ is denoted by $\Phi(z_0)$, so

$$\Phi(z_0) = P(Z \le z_0) = \frac{1}{\sqrt{2\pi}} \int_{-\infty}^{z_0} e^{-t^2/2} \, dt.$$

Table B.1. Standard Normal Cumulative Probability in Left-Hand Tail

z_0	0.00	0.01	0.02	0.03	0.04	0.05	0.06	0.07	0.08	0.09
0.0	0.5000	0.5040	0.5080	0.5120	0.5160	0.5199	0.5239	0.5279	0.5319	0.5359
0.1	0.5398	0.5438	0.5478	0.5517	0.5557	0.5596	0.5636	0.5675	0.5714	0.5753
0.2	0.5793	0.5832	0.5871	0.5910	0.5948	0.5987	0.6026	0.6064	0.6103	0.6141
0.3	0.6179	0.6217	0.6255	0.6293	0.6331	0.6368	0.6406	0.6443	0.6480	0.6517
0.4	0.6554	0.6591	0.6628	0.6664	0.6700	0.6736	0.6772	0.6808	0.6844	0.6879
0.5	0.6915	0.6950	0.6985	0.7019	0.7054	0.7088	0.7123	0.7157	0.7190	0.7224
0.6	0.7257	0.7291	0.7324	0.7357	0.7389	0.7422	0.7454	0.7486	0.7517	0.7549
0.7	0.7580	0.7611	0.7642	0.7673	0.7704	0.7734	0.7764	0.7794	0.7823	0.7852
0.8	0.7881	0.7910	0.7939	0.7967	0.7995	0.8023	0.8051	0.8078	0.8106	0.8133
0.9	0.8159	0.8186	0.8212	0.8238	0.8264	0.8289	0.8315	0.8340	0.8365	0.8389
1.0	0.8413	0.8438	0.8461	0.8485	0.8508	0.8531	0.8554	0.8577	0.8599	0.8621
1.1	0.8643	0.8665	0.8686	0.8708	0.8729	0.8749	0.8770	0.8790	0.8810	0.8830
1.2	0.8849	0.8869	0.8888	0.8907	0.8925	0.8944	0.8962	0.8980	0.8997	0.9015
1.3	0.9032	0.9049	0.9066	0.9082	0.9099	0.9115	0.9131	0.9147	0.9162	0.9177
1.4	0.9192	0.9207	0.9222	0.9236	0.9251	0.9265	0.9279	0.9292	0.9306	0.9319
1.5	0.9332	0.9345	0.9357	0.9370	0.9382	0.9394	0.9406	0.9418	0.9429	0.9441
1.6	0.9452	0.9463	0.9474	0.9484	0.9495	0.9505	0.9515	0.9525	0.9535	0.9545
1.7	0.9554	0.9564	0.9573	0.9582	0.9591	0.9599	0.9608	0.9616	0.9625	0.9633
1.8	0.9641	0.9649	0.9656	0.9664	0.9671	0.9678	0.9686	0.9693	0.9699	0.9706
1.9	0.9713	0.9719	0.9726	0.9732	0.9738	0.9744	0.9750	0.9756	0.9761	0.9767
2.0	0.9772	0.9778	0.9783	0.9788	0.9793	0.9798	0.9803	0.9808	0.9812	0.9817
2.1	0.9821	0.9826	0.9830	0.9834	0.9838	0.9842	0.9846	0.9850	0.9854	0.9857
2.2	0.9861	0.9864	0.9868	0.9871	0.9875	0.9878	0.9881	0.9884	0.9887	0.9890
2.3	0.9893	0.9896	0.9898	0.9901	0.9904	0.9906	0.9909	0.9911	0.9913	0.9916
2.4	0.9918	0.9920	0.9922	0.9925	0.9927	0.9929	0.9931	0.9932	0.9934	0.9936
2.5	0.9938	0.9940	0.9941	0.9943	0.9945	0.9946	0.9948	0.9949	0.9951	0.9952
2.6	0.9953	0.9955	0.9956	0.9957	0.9959	0.9960	0.9961	0.9962	0.9963	0.9964
2.7	0.9965	0.9966	0.9967	0.9968	0.9969	0.9970	0.9971	0.9972	0.9973	0.9974
2.8	0.9974	0.9975	0.9976	0.9977	0.9977	0.9978	0.9979	0.9979	0.9980	0.9981
2.9	0.9981	0.9982	0.9982	0.9983	0.9984	0.9984	0.9985	0.9985	0.9986	0.9986

The following example illustrates the use of this table.

Example B.4. Suppose that the daily price change X of an investment is normally distributed with mean $\mu = 0.05$ and variance $\sigma^2 = 0.01$. For a given day, what is the probability that the price change is

(a) Less than 0.05?
(b) Greater than −0.05?
(c) Greater than 0.15?
(d) Between −0.05 and 0.15?

Solution. In the following we let $Z = aX + b$, where $a = 1/0.10$ and $b = -0.05/0.10$. It is easy to see that Z has the standard normal distribution. In some of these calculations we make use of the fact that the distribution function of Z is symmetric about 0, that is, $P(Z \geq z) = P(Z \leq -z)$ for any real number z.

(a) We have

$$P(X < 0.05) = P((X - 0.05)/0.10 < (0.05 - 0.05)/0.10)$$
$$= P(Z < 0)$$
$$= 0.5000.$$

(b) We have

$$P(X > -0.05) = P((X - 0.05)/0.10 > (-0.05 - 0.05)/0.10)$$
$$= P(Z > -1)$$
$$= P(Z \leq 1)$$
$$= 0.8413.$$

(c) We have

$$P(X > 0.15) = P(Z > (0.15 - 0.05)/0.10)$$
$$= P(Z > 1)$$
$$= 1 - P(Z \leq 1)$$
$$= 0.1587.$$

(d) We have

$$P(-0.05 < X < 0.15) = P(-1 < Z < 1)$$
$$= P(Z < 1) - P(Z \leq -1)$$
$$= 0.8413 - 0.1587$$
$$= 0.6826.$$

△

If X_i, $i = 1, 2, \ldots, n$, are independent normal random variables, and if X_i has mean μ_i and variance σ_i^2 for each i, then it can be shown that $\sum_{i=1}^{n} X_i$ has a normal distribution with mean $\sum_{i=1}^{n} \mu_i$ and variance $\sum_{i=1}^{n} \sigma_i^2$.

We have the following theorem, which we state without proof.[3]

Theorem B.4. *The Central Limit Theorem.*

If X_1, X_2,... are independent random variables that have a common distribution function with mean μ, $-\infty < \mu < \infty$, and variance σ^2, $0 < \sigma^2 < \infty$, then, for any real number z,

$$\lim_{n \to \infty} P\left(\frac{\frac{1}{n}\sum_{i=1}^{n} X_i - \mu}{\sigma/\sqrt{n}} \le z\right) = \Phi(z),$$

where $\Phi(z) = P(Z \le z)$, the distribution function of the standard normal random variable Z, evaluated at z.

We write this as

$$Z_n = \frac{\frac{1}{n}\sum_{i=1}^{n} X_i - \mu}{\sigma/\sqrt{n}}$$

converges in distribution to the standard normal distribution[4] and denote this by $Z_n \xrightarrow{d} Z$. Notice that the restrictions placed on the random variables $X_1, X_2,...$ are that they are independent, have the same distribution, and have a finite and non-zero variance.[5]

Example B.5. Suppose that we toss a fair coin a large number of times. We define the random variables X_1, X_2,... by $X_i = 1$ if the ith toss is a head and $X_i = 0$ if the ith toss is a tail. Each random variable X_i has a binomial distribution with parameters $n = 1$ and $p = 0.50$. If we assume that the tosses are independent and that the coin is tossed n times, then $X = \sum_{i=1}^{n} X_i$ has a binomial distribution with mean $0.5n$ and variance $n \times 0.5 \times 0.5 = 0.25n$. Answer the following questions:

(a) If the coin is tossed 100 times, then what is $P(X = 50)$?
(b) If the coin is tossed 100 times, then what is $P(X \ge 60)$?
(c) If the coin is tossed 100 times, then what is $P(40 \le X \le 60)$?

Solution. We can approximate these probabilities using the Central Limit Theorem. We use the fact that with $n = 100$ and $p = 0.50$, the random variable X has a binomial distribution with mean $\mu = E(X) = np = 50$ and variance $\sigma^2 = np(1 - p) = 100 \times 0.50 \times 0.50 = 25$. Thus, $(\sum_{i=1}^{100} X_i - 50)/5$ has an approximate standard normal distribution. Note that because X is a discrete random variable and Z is a continuous random variable, events such as $(X = 50)$ need to be rewritten as $(49.5 < X < 50.5)$ in order to make use of the Central Limit Theorem.

---

[3] A proof of the Central Limit Theorem can be found in [17].
[4] Let X_1, X_2,... and Y be random variables with distribution functions Φ_1, Φ_2,... and ψ, respectively. We say that X_n converges in distribution to Y, written $X_n \xrightarrow{d} Y$, if $\Phi_n(x) \to \psi(x)$ at each continuity point of ψ.
[5] If a random variable has a finite variance, then it also has a finite mean.

(a) We have

$$P(X = 50) = P(49.5 < X < 50.5)$$
$$= P((49.5 - 50)/5 < (X - 50)/5 < (50.5 - 50)/5)$$
$$\approx P(-0.1 < Z < 0.1)$$
$$= P(Z \leq 0.1) - P(Z \leq -0.1)$$
$$= 0.5398 - 0.4602$$
$$= 0.0796.$$

(b) We have

$$P(X \geq 60) = P(X \geq 59.5)$$
$$= P((X - 50)/5 \geq (59.5 - 50)/5)$$
$$\approx P(Z \geq 1.90)$$
$$= 1 - P(Z \leq 1.90)$$
$$= 1 - 0.9713$$
$$= 0.0287.$$

(c) We have

$$P(40 \leq X \leq 60) = P(39.5 \leq X \leq 60.5)$$
$$= P((39.5 - 50)/5 \leq (X - 50)/5 \leq (60.5 - 50)/5)$$
$$\approx P(-2.10 \leq Z \leq 2.10)$$
$$= P(Z \leq 2.10) - P(Z \leq -2.10)$$
$$= 0.9821 - 0.0179$$
$$= 0.9642.$$

$\triangle$

Another important continuous random variable is the LOGNORMAL RAN-DOM VARIABLE. A random variable X is said to be lognormal if $Y = \ln X$ has a normal distribution. Note that X can only take positive values. Lognormal distributions are often used to model rates of returns on investments. Suppose that $Y = \ln X$ has a normal distribution with mean μ and variance σ^2. Note that $f(x) = 0$ for $x \leq 0$ and that $f(x) > 0$ for $x > 0$. It is left as an exercise to derive the density function.

B.4 Moments

Although the Central Limit Theorem works very well in many cases, there are instances where one or more of its assumptions are not met. The conditions on the mean and variance ($-\infty < \mu < \infty$, $0 < \sigma^2 < \infty$) are usually satisfied in real-world cases, but the other conditions may not be satisfied. For example, suppose that for each n, X_i depends on n. Another example is the case where $X_1, X_2, \ldots, X_n$ are not independent random variables. In both cases the Central Limit Theorem is not applicable, but it may still be possible that $Z_n = (X_1 + \cdots + X_n - n\mu)/(\sigma\sqrt{n})$ converges in distribution to Z, the standard normal random variable.

To explore these cases further we introduce the concept of moment generating functions. Let X be a random variable and t a real number. The MOMENT GENERATING FUNCTION of the random variable X is defined by $M_X(t) = E(e^{tX})$ for all t for which the expectation exists. Note that $M_X(0)$ exists for any random variable, but there are random variables for which $M_X(t)$ does not exist for $t \neq 0$.

Consider the following two cases.

Case 1. Let X be a binomial random variable with $n = 1$ and probability of success p. Then

$$M_X(t) = E(e^{tX}) = pe^{t \times 1} + (1-p)e^{t \times 0} = pe^t + (1-p).$$

Case 2. Let X have a normal distribution with mean μ and variance σ^2. Then

$$
\begin{aligned}
M_X(t) &= E(e^{tX}) \\
&= \frac{1}{\sqrt{2\pi\sigma^2}} \int_{-\infty}^{\infty} e^{tx} e^{-(x-\mu)^2/(2\sigma^2)} \, dx \\
&= \frac{1}{\sqrt{2\pi\sigma^2}} \int_{-\infty}^{\infty} e^{tx - (x-\mu)^2/(2\sigma^2)} \, dx.
\end{aligned}
$$

We have

$$
\begin{aligned}
tx - \frac{(x-\mu)^2}{2\sigma^2} &= \frac{1}{2\sigma^2}\left(2\sigma^2 tx - (x-\mu)^2\right) \\
&= \frac{1}{2\sigma^2}\left(-\left(x - (\sigma^2 t + \mu)\right)^2 + \sigma^4 t^2 + 2\mu\sigma^2 t\right).
\end{aligned}
$$

So

$$M_X(t) = e^{\mu t + \sigma^2 t^2/2} \frac{1}{\sqrt{2\pi\sigma^2}} \int_{-\infty}^{\infty} e^{-\left(x - (\sigma^2 t + \mu)\right)^2/(2\sigma^2)} \, dx.$$

We introduce

$$v = \frac{x - (\sigma^2 t + \mu)}{\sigma},$$

to obtain

$$
\begin{aligned}
M_X(t) &= e^{\mu t + \sigma^2 t^2/2} \frac{1}{\sqrt{2\pi}} \int_{-\infty}^{\infty} e^{-v^2/2} \, dv \\
&= e^{\mu t + \sigma^2 t^2/2}.
\end{aligned}
$$

In the important case where $X = Z$, the standard normal random variable, we have $M_X(t) = e^{t^2/2}$.

The following theorem, which we state without proof,[6] is often useful in deriving limiting distributions of random variables in cases where the Central Limit Theorem is not applicable.

Theorem B.5. *Moment Generating Function Convergence Theorem.*
Let $Z_1, Z_2, \ldots$ be random variables such that $M_{Z_n}(t)$ exists for every real number t and all $n \geq 1$. Then if $M_{Z_n}(t)$ converges to $M_T(t)$ for every real number t, where T is a random variable, then $P(Z_n \leq x)$ converges to $P(T \leq x)$ for every real number x. Thus, if the moment generating functions converge, then the distribution functions converge and $Z_n \xrightarrow{d} T$.

Using the Moment Generating Function Convergence Theorem we may prove the following useful result.

Theorem B.6. *Binomial Convergence Theorem.*
For every n, let $S_n = X_1 + X_2 + \cdots + X_n$, where $X_1, X_2, \ldots, X_n$ are independent with $P(X_i = 1) = p_n$ and $P(X_i = 0) = 1 - p_n$ for $i = 1, 2, \ldots, n$. This has a binomial distribution with parameters n and p_n. If $Z_n = (S_n - np_n)/\sqrt{np_n(1 - p_n)}$ and if p_n converges to p as n approaches ∞, with $0 < p < 1$, then $Z_n \xrightarrow{d} Z$.

Proof. We use the Moment Generating Function Convergence Theorem.

We have $P(X_i = 1) = p_n$ and $P(X_i = 0) = 1 - p_n$ for $i = 1, 2, \ldots, n$. It is important to note that for each n the X_i's may depend on n.

Then

$$M_{Z_n}(t) = E(e^{tZ_n})$$
$$= E\left(e^{t(S_n - np_n)/\sqrt{np_n(1-p_n)}}\right)$$
$$= E\left(e^{t(X_1 + X_2 + \cdots + X_n - np_n)/\sqrt{np_n(1-p_n)}}\right)$$
$$= E\left(e^{t(X_1 - p_n)/\sqrt{np_n(1-p_n)}}\right) \cdots E\left(e^{t(X_n - p_n)/\sqrt{np_n(1-p_n)}}\right)$$
$$= \left(E\left(e^{t(X_1 - p_n)/\sqrt{np_n(1-p_n)}}\right)\right)^n,$$

where we used the facts that the X_i's are independent and identically distributed for each $n \geq 1$.

It remains to show that

$$\lim_{n \to \infty} M_{Z_n}(t) = M_Z(t) = e^{t^2/2}$$

for all t.

---

[6] A proof of the Moment Generating Function Convergence Theorem can be found in [8].

Expanding $e^{t(X_1-p_n)/\sqrt{np_n(1-p_n)}}$ in a Taylor series about $t = 0$, we see that

$$e^{t(X_1-p_n)/\sqrt{np_n(1-p_n)}} = 1 + \frac{t(X_1-p_n)}{\sqrt{np_n(1-p_n)}} + \frac{t^2(X_1-p_n)^2}{2!np_n(1-p_n)} + \cdots.$$

Now, $E(X_1 - p_n) = 0$ and $E(X_1 - p_n)^2 = \sigma_{X_1}^2 = p_n(1-p_n)$, so

$$E\left(1 + \frac{t(X_1-p_n)}{\sqrt{np_n(1-p_n)}} + \frac{t^2(X_1-p_n)^2}{2np_n(1-p_n)}\right) = 1 + \frac{t^2}{2n}$$

for every n. Thus, we have

$$M_{Z_n}(t) = \left(1 + \frac{t^2}{2n}(1 + r_n)\right)^n,$$

where

$$r_n = 2\sum_{k=3}^{\infty} \frac{t^{k-2}n^{1-k/2}}{k!} E\left(\left(\frac{X_1-p_n}{\sqrt{p_n(1-p_n)}}\right)^k\right).$$

Now,

$$\left|\frac{X_1-p_n}{\sqrt{p_n(1-p_n)}}\right| = \begin{cases} \sqrt{\frac{1-p_n}{p_n}} & \text{if } X_1 = 1, \\ \sqrt{\frac{p_n}{1-p_n}} & \text{if } X_1 = 0. \end{cases}$$

Thus, for large enough n,

$$\left|\frac{X_1-p_n}{\sqrt{p_n(1-p_n)}}\right| \le cM,$$

where c is any number greater than 1, and

$$M = \max\left(\sqrt{\frac{1-p}{p}}, \sqrt{\frac{p}{1-p}}\right),$$

so

$$\left|\frac{t^{k-2}n^{1-k/2}}{k!} E\left(\left(\frac{X_1-p_n}{\sqrt{p_n(1-p_n)}}\right)^k\right)\right| \le \frac{t^{k-2}n^{1-k/2}}{k!}(cM)^k$$

$$= t^{k-2}n^{1-k/2}\frac{(cM)^k}{k!}.$$

From this we see that r_n approaches 0 as n approaches infinity. Thus, it follows (see Problem B.20 on p. 282) that

$$\lim_{n\to\infty} M_{Z_n}(t) = M_Z(t) = e^{t^2/2}.$$

This proves that

$$Z_n \xrightarrow{d} Z,$$

by the Moment Generating Function Convergence Theorem.   □

B.5 Joint Distribution of Random Variables

Suppose that X and Y are two discrete random variables defined on the same sample space. Then we may determine the joint distribution of X and Y from their joint probability function as follows: For all real numbers x and y, define $f(x, y) = P(X = x, Y = y)$. Consider the following joint probability distribution of the random variables X and Y.

$X \backslash Y$	-1	0	1
0	1/12	1/8	1/24
1	1/6	1/4	1/12
2	1/12	1/8	1/24

Thus, for example, $f(1, 0) = P(X = 1, Y = 0) = 1/4$ and $f(0, 1) = P(X = 0, Y = 1) = 1/24$. Note that we may calculate the probability function of X as follows:

$$P(X = 0) = f(0, -1) + f(0, 0) + f(0, 1) = \frac{1}{4},$$

$$P(X = 1) = f(1, -1) + f(1, 0) + f(1, 1) = \frac{1}{2},$$

and, similarly,

$$P(X = 2) = \frac{1}{4}.$$

From these we may calculate $E(X) = 1$ and $\sigma_X^2 = 0.5$.

Using this procedure, we may calculate the probability function of Y, and thus the parameters $E(Y)$ and σ_Y^2. We may also find the probability function of the random variable XY, which is given by $P(XY = -2) = 1/12$, $P(XY = -1) = 1/6$, $P(XY = 0) = 5/8$, $P(XY = 1) = 1/12$, and $P(XY = 2) = 1/24$. From these we may calculate $E(XY)$ and σ_{XY}^2. You should do this.

If we plot these data, then we have Fig. B.5. There does not appear to be a strong linear relationship between X and Y.

On the other hand, consider the following joint probability distribution of the random variable S and T.

$S \backslash T$	-1	0	1
0	1/6	1/8	1/24
1	1/12	1/8	1/12
2	1/24	1/12	1/4

If we plot these data, then we have Fig. B.6.

In this case there appears to be a positive linear relationship between S and T. For example, $f(0, -1) = P(S = 0, T = -1) = 1/6$, whereas $f(0, 1) = P(S = 0, T = 1) = 1/24$. Also, $f(2, 1) = 1/4$, and $f(2, -1) = 1/24$. Larger values of S are more likely to occur with larger values of T, and smaller values of S are more likely to occur with smaller values of T.

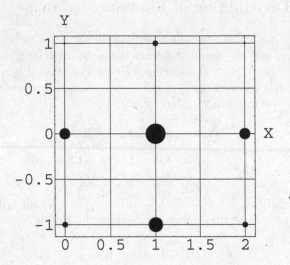

Fig. B.5. Joint probability function of X and Y

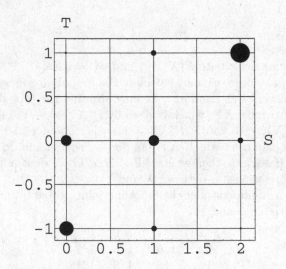

Fig. B.6. Joint probability function of S and T

A measure of linearity is given by the covariance, which is defined for random variables X and Y by

$$\sigma_{XY} = E\left((X - E(X))(Y - E(Y))\right) = E(XY) - E(X)E(Y).$$

For the random variables X and Y given at the beginning of this section, $\sigma_{XY} = 0$, whereas for the random variables S and T, $\sigma_{ST} = 0.3299$. However, this measure of linearity is dependent on the units of measurement. For example, it is easy to show that $\sigma_{(2S)T} = 2\sigma_{ST} = 0.6597$. A unitless measure of linearity is the correlation coefficient, which is defined for random variables X and Y by

$$\rho_{XY} = \frac{\sigma_{XY}}{\sqrt{\sigma_X^2 \sigma_Y^2}}.$$

Using this measure, $\rho_{XY} = 0$ and $\rho_{ST} = 0.4831$. By Hölder's Inequality, we have $-1 \leq \rho_{XY} \leq 1$. We have $\rho_{XY} = 1$ if and only if $Y = a + bX$ with $b > 0$, and we have $\rho_{XY} = -1$ if and only if $Y = c + dX$ with $d < 0$. (See Problem B.17.)

B.6 Linear Regression

Consider the following data collected on the relationship between family size and annual family income.

Number of Children in Family, X	Annual Family Income (Thousands of Dollars), Y
1	55
2	47
3	45
4	43
5	40
6	38

If we plot these data, then we have Fig. B.7.

We wish to fit a line—called a REGRESSION LINE—to these data that describes the relationship between X and Y. How to "best" fit the line is the subject of this section. Consider the general case where we have n pairs of points: $(x_1, y_1), (x_2, y_2), \ldots, (x_n, y_n)$. One possible criterion for fitting a line to the data is to choose a line $\hat{y} = a + bx$, which minimizes the sum of the absolute values of the errors. That is, we seek a line that minimizes $\sum_{i=1}^{n} |y_i - (a + bx_i)|$. In some cases this is a reasonable criterion, but it has several deficiencies.

For example, consider the following data set, plotted in Fig. B.8.

x	y
1	1
2	2
3	2
4	1

Now, consider three possible lines: L1: $y = 1.0$ ($a = 1, b = 0$), L2: $y = 1.5$ ($a = 1.5, b = 0$), and L3: $y = 2.0$ ($a = 2, b = 0$). It is easy to see that all three lines yield the same sum of absolute values of errors, namely 2. Thus, there is

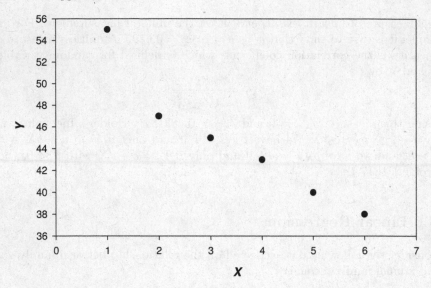

Fig. B.7. Annual family income versus number of children in family

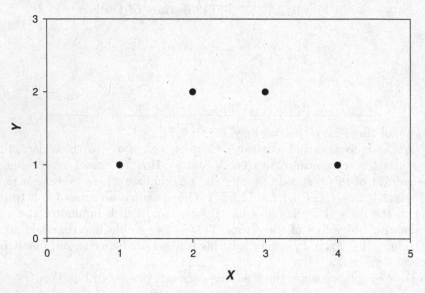

Fig. B.8. The data set

no way to choose between the three lines using this criterion. In general, there may be more than one line that minimizes the sum of the absolute values of the errors. Also, this method of choosing the "best fit" line is usually not amenable to statistical analysis.

The most common method of fitting lines to bivariate data sets uses the "least-squares" criterion. With this criterion we choose the line $\widehat{y} = a + bx$ that minimizes the sum of the squares of the errors, $\sum_{i=1}^{n} (y_i - (a + bx_i))^2$. This method is preferred because it yields a unique line and leads to useful applied and theoretical statistical applications. We now derive the equations for a and b that give the LEAST-SQUARES REGRESSION LINE.

We are given n pairs of points: $(x_1, y_1), (x_2, y_2), \ldots, (x_n, y_n)$, and we are to find the line $\widehat{y} = a + bx$ that minimizes $\sum_{i=1}^{n} (y_i - (a + bx_i))^2$. We derive the equations for a and b using calculus. We first differentiate the expression $\sum_{i=1}^{n} (y_i - (a + bx_i))^2$ with respect to a and set the derivative equal to zero. Then differentiate the (original) expression with respect to b and set the derivative equal to zero. This leads to two linear equations in a and b, which we solve for the desired values.

First, differentiating with respect to a, we have

$$-2 \sum_{i=1}^{n} (y_i - (a + bx_i)) = 0,$$

so

$$\sum_{i=1}^{n} y_i = na + b \sum_{i=1}^{n} x_i.$$

Second, differentiating with respect to b, we have

$$-2 \sum_{i=1}^{n} x_i (y_i - (a + bx_i)) = 0,$$

so

$$\sum_{i=1}^{n} x_i y_i = a \sum_{i=1}^{n} x_i + b \sum_{i=1}^{n} x_i^2.$$

If we set $\overline{x} = \sum_{i=1}^{n} x_i / n$ and $\overline{y} = \sum_{i=1}^{n} y_i / n$ and solve for a and b, then we have

$$a = \overline{y} - b\overline{x},$$

and

$$b = \frac{\sum_{i=1}^{n} (x_i - \overline{x})(y_i - \overline{y})}{\sum_{i=1}^{n} (x_i - \overline{x})^2}$$

The least squares regression line is $y = a + bx$.

Referring back to the data set plotted in Fig. B.8, it is easy to see that L2: $y = 1.5$ is the least-squares regression line. The calculation of the least-squares regression line for family size versus annual family income is left as an exercise. For that example, should b be less than, equal to, or greater than zero? From Fig. B.7, estimate the value of a.

B.7 Estimates of Parameters of Random Variables

Let X and Y be random variables, with means $E(X)$ and $E(Y)$, variances σ_X^2 and σ_Y^2, covariance σ_{XY}, and correlation coefficient ρ_{XY}. In real-world applications these parameters are usually unknown, and they must be estimated from random samples.

Example B.6. Consider the following sample of verbal test scores and high school grade point averages for first-year students at a university. Use this sample to estimate the following parameters: $E(X)$, $E(Y)$, σ_Y^2, σ_X^2, σ_{XY}, and ρ_{XY}.

Verbal Test Score, X	High School GPA, Y
35	2.73
50	2.93
62	3.25
58	3.30
75	3.49

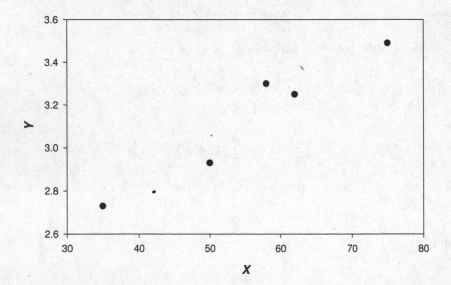

Solution. In general, we have a bivariate random sample $(x_1, y_1), (x_2, y_2), \ldots, (x_n, y_n)$ from the population of all pairs (X, Y).

1. The estimate of $E(X)$ is $\overline{x} = \sum_{i=1}^{n} x_i / n$, the sample mean. In this example, $\overline{x} = 56$, the arithmetic average of $x_1, x_2, \ldots, x_5$. The estimate of $E(Y)$ is $\overline{y} = 3.14$.

2. The estimate of σ_X^2 is $s_X^2 = \sum_{i=1}^n (x_i - \overline{x})^2/(n-1)$, the sample variance. In this example, the estimate of σ_X^2 is $s_X^2 = 219.50$, and the estimate of σ_Y^2 is $s_Y^2 = 0.0931$.
3. The estimate of σ_{XY} is $s_{XY} = \sum_{i=1}^n (x_i - \overline{x})(y_i - \overline{y})/(n-1)$. In this example, the estimate of σ_{XY} is $s_{XY} = 4.375$.
4. The estimate of ρ_{XY} is $r_{XY} = s_{XY}/\sqrt{s_X^2 s_Y^2}$, the sample correlation coefficient. In this example, the estimate of ρ_{XY} is $r_{XY} = 0.9678$. The estimator, r, satisfies the same restriction as the population correlation coefficient, ρ, namely $-1 \le r \le 1$. Note that $r_{XY} = 0.9678$ is very close to 1, indicating that there is a strong positive linear relationship between X and Y. This should be obvious by looking at the figure.

$\triangle$

B.8 Problems

B.1. If X is a random variable and $(X \subset A)$ is an event, then the **indicator variable** $1_{(X \subset A)}$ is defined to be 1 if $X \subset A$, and 0 if $X \not\subset A$.

(a) Show that $E(1_{(X \subset A)}) = P(X \subset A)$.
(b) If $A_1, A_2, \ldots, A_n$ are disjoint, then what is $E(1_{(X \subset A_1)} + 1_{(X \subset A_2)} + \cdots + 1_{(X \subset A_n)})$?
(c) If X_1 and X_2 are independent random variables and x_1, x_2 are real numbers, then what is $E(1_{(X_1 < x_1)} 1_{(X_2 < x_2)})$?

B.2. Verify that the variance of a binomial random variable is $np(1-p)$.

B.3. Verify the values of $f(k)$ and $g(k)$ on p. 264.

B.4. Verify that $d_1 = d_2 = 0.9513$ on p. 264.

B.5. Let $\lambda > 0$ and

$$ f(x) = \begin{cases} c\frac{\lambda^x}{x!}, & \text{if } x = 0, 1, \ldots, \\ 0 & \text{otherwise.} \end{cases} $$

(a) For what constant c is $f(x)$ a probability function?
(b) What is $E(X)$?
(c) What is σ^2?

B.6. Let X be any random variable, with moment-generating function $M_X(t)$. What is the value of $M_X(0)$?

B.7. Verify the calculations for σ_{XY} and σ_{ST} on p. 276.

B.8. Prove that $\sum_{i=1}^n (x_i - \overline{x}) = 0$.

B.9. Prove that $\sum_{i=1}^{n}(x_i - \bar{x})^2 = \sum_{i=1}^{n} x_i^2 - \left(\sum_{i=1}^{n} x_i\right)^2 / n$.

B.10. Compute the correlation coefficient between the DJIA and NASDAQ indexes from 1980 through 2005.

B.11. In deriving the coefficients a and b for the least-squares regression line, we differentiated $\sum_{i=1}^{n}(y_i - (a + bx_i))^2$ with respect to a and with respect to b and set the derivatives equal to zero to obtain the desired intercept and slope for the line. Is this enough to guarantee that these values minimize the sum of the squared errors? If not, complete the proof.

B.12. Suppose that $(x_1, y_1), (x_2, y_2), \ldots, (x_n, y_n)$ are given, and we wish to minimize $\sum_{i=1}^{n}\left(y_i - \left(c + dx_i + ex_i^2\right)\right)^2$. That is, we wish to find the quadratic function $\hat{y} = c + dx + ex^2$ that minimizes the sum of the squared errors. Using calculus, derive the proper values of the constants c, d, and e.

B.13. Verify the values of the sample statistics on p. 280.

B.14. Let X have a normal distribution with density function

$$\frac{1}{\sqrt{2\pi\sigma^2}}\, e^{-(x-\mu)^2/(2\sigma^2)}.$$

Show that $\sigma^2 = E\left((X - \mu)^2\right)$. [Hint: Use integration by parts.]

B.15. Let X have a lognormal distribution, with $Y = \ln X$ having a normal distribution with mean μ and variance σ^2. Derive the density function of X.

B.16. Prove that $\sum_{i=1}^{n} x_i^2 \geq \left(\sum_{i=1}^{n} x_i\right)^2 / n$ using the Cauchy-Schwarz Inequality.

B.17. Prove that $-1 \leq r_{XY} \leq 1$ using Hölder's Inequality.

B.18. Calculate the sample mean and sample variance for the DJIA data from 1980 through 2005.

B.19. For any real number x, define $[x]^+$ by

$$[x]^+ = \begin{cases} x & \text{if } x > 0, \\ 0 & \text{if } x \leq 0. \end{cases}$$

(a) When is $[ax]^+ = ax$?
(b) Is $[x]^+ + [y]^+ = [x + y]^+$? Prove this or give a counter-example. If it is false, under what conditions is it true?

B.20. For $n \geq 1$, let $r(n)$ be a function of n converging to 0 as n approaches infinity. Show that $(1 + (x + r(n))/n)^n$ converges to e^x as n approaches infinity for any x. [Hint: Use the fact that $(1 + (y/n))^n$ converges to e^y for any y.]

Answers

1.1 $0.075 = 7.5\%$

1.2 $0.075 = 7.5\%$; same

1.3 $3.29

1.4 Withdrawal made before 90 days

1.5 212 days

1.6 210 days

1.9 Oct. 4, 1998 and Oct. 31, 1998

1.10 No

2.2 $0.0678 = 6.78\%$; simple interest

2.3 $28.84

2.4 $0.099 = 9.9\%$; $0.1041 = 10.41\%$

2.5 5.893 years; $0.071225 = 7.1\%$

2.6 $0.1487 = 14.87\%$

2.7 $6727.50

2.8 IRR $= 0$

2.9 0.1487, 0.1583, 0.1161

2.10 $11.798 = 1179.8\%$

2.11 $315,241.71

2.12 Same

2.13 $0.0772 = 7.72\%$

2.14 Concave down

2.16 7%, 8%, 9%, after rounding

2.17 $8, 9, 10, 11$ years

2.18 $0.0639 = 6.39\%$

2.19 $3,980.11

2.20 $x = -1.32, 1, 1.034$

2.21

(a) 0.06015
(b) 0.0609
(c) 0.0618

2.22

(a) $593.51
(b) $660.76
(c) $463.51
(d) $537.71

2.23

(a) 0.0497
(b) 0.05016
(c) 0.05017
(d) 0.0503

2.24

(a) 0.0719
(b) 0.071
(c) 0.0714

(a) is highest

2.25 $0.1712 = 17.12\%$

2.43 No

2.45 $n = 1, C(0) = -1, C(1) = -1$

2.49 $0.0785 = 7.85\%$

3.2 1921, 1922, 1927, 1928, 1930, 1931, 1932, 1933, 1938, 1939, 1949, 1955

3.3 0.03512, 0.027137, 0.02550

3.4 No; -0.009

3.5 $253.98

3.6 $0.0670 = 6.7\%$

3.12 $(i_{\text{eff}}(1 - t) - i_{\text{inf}})/(1 + i_{\text{inf}})$

3.14 $f(x) = x$ if $x \leq 1$,
$f(x) = 2 - x$ if $x > 1$

4.3 Yes, if compounded monthly

4.4 $14,285.71

4.5 $25,750.00

4.6 $7,506.13

4.7 $9,675.71

4.8 $9,748.28

4.9 $1,450.53

4.10 $1,313.04

4.11 $245.92

4.12 $10,222.58

4.13 $0.0814 = 8.14\%$.

4.14 $0.104 = 10.4\%$

4.15 $5,841.18

4.16 $5,408.78

4.17 $334.25

4.24 Yes

5.3 (a) $16,377.98

5.6 Yes; Yes (except for rounding)

5.7 (a) $0.05 = 5\%$

6.6 $0.025 = 2.5\%$

6.7 No

6.8 $394.91, exact

6.9 13.22, 12.68, 12.15, 11.62, 11.08
132.16, 126.83, 121.50, 116.16, 110.83
145.38, 139.51, 133.64, 127.78, 121.91
$0.10 = 10\%$
estimates too large for years 11, 12, 13

6.11 No, paid off in less than half the time

6.14 $132.16, $118.71, $90.88, $73.90, $264.31, $237.41, $181.75, $147.80

7.2 226 months (18 yr 10 months), $4,786.66; 58 months (4 yr 10 months), $3,141.25

7.5 No, 6

7.6 Yes

7.7 Yes

7.11 Less total interest paid

7.12 Increases

7.14 Yes

7.15 Yes, unless the digit is 9

8.9 Hugh: Bond 1: 0.0506,
Bond 2: 0.1025, Port.: 0.0764
Tom: Bond 3: 0.0404, Bond 4: 0.09203,
Port.: 0.0816.

8.10 $3,000.00

8.11 $0.0975 = 9.75\%$

8.12

(a) $12,940.07
(b) $0.08 = 8\%$

8.13 Wendy: 9.313 years, 9.042 years
Amanda: 7.286 years, 7.074 years

8.14 8.360 years, 8.117 years

8.22

(a) 2
(b) $2/y^2$
(c) $2/(1 + 4y^2)^{3/2}$

No; (a); no

8.23

(a) e^y
(b) 1
(c) $e^y/(1 + e^{2y})^{3/2}$

No; (b); no

8.24

(a) $-1/(1 - y^2)^{3/2}$
(b) $-1/(1 - y^2)^2$
(c) -1

Yes; (c); yes

9.1 AT&T: T $26.03
International Paper: IP $34.00
Verizon: VZ $33.48

9.2 (a) March 13, 1986

9.3 Hugh

9.5 Below $20 a share

9.6 $34.62 per share

9.9 Profit: Tom $3300,
Wendy $2533.33
Wendy greater profit per share.
Tom greater total profit.

9.10 27.273 shares

10.1

(a) 31.67, 33.67, 34.67, 36.00, 34.67
(b) 10, 13.66, 14.05, 16.22, 15.51
(c) DJIA: Arith. 0.02357,
Geom. 0.02289;
S&P 500: Arith. 0.1263,
Geom. 0.1159

10.2

(a) -0.005, 0.046, 0.02, 0.06
(b) Arith. 0.01; Geom. -0.00134;
Arithmetic-Geometric Mean Inequality

10.3 0.1409

10.4 0.3460

10.5 101.3203

10.6 Increases

10.8 Doubles

10.10 200: product relative prices: 16
50: product relative prices: 1/16

10.12 DJIA: 1011.79\%,
S&P 500: 819.48\%,
NASDAQ: 989.91\%

10.13 $S(t) = 2^t + 1, D_t = 1, k = 1$

11.11 $4.08

11.12 No. Buy call option for $8,
buy zero-coupon bond with face value
of $100 for $e^{-0.05}100 = \$95.12$, short
stock for $105, short put for $2. Profit
$3.88

11.14 $S(t_0) + P = C + (X + D)e^{-i(\infty)T}$

11.16 $\Delta^* = -2/3, B = 200/3$

11.20 96.8096

11.22

(a) 206.076
(b) 212.336
(c) 0.0599
(d) 0.0617
(e) 0.06076

11.23

(a) $(u + 1/u)S(t_0)/2$
(b) $S(t_0)$

11.24

(a) $(pu)^n S(t_0)$
(b) $0, S(t_0), \infty$
(c) $pu < 1$

11.26 $15, 1.32, 0, 0, 11.46, 0.46,$
$0, 6.048, 0.281, 1.896$

11.28 Some $p > 1$

11.42 13.59

11.43 11.68

11.44 12.36

11.45 7.46

11.46 13.95

A.7 No; $f(41) = 41 \cdot 43$

B.1

(b) $P[X \subset A_1] + \cdots + P[X \subset A_n]$
(c) $P[X_1 < x_1]P[X_2 < x_2]$

B.5

(a) $c = e^{-\lambda}$
(b) λ
(c) λ

B.6 1

B.10 0.9371

B.18 Mean = 5138.804,
variance = 14862318.404

B.19

(a) $ax \geq 0$
(b) Counter example: $x = 1, y = -1$.
     True when $x \geq 0$ and $y \geq 0$, or
     when $x \leq 0$ and $y \leq 0$.

References

1. Abramowitz, M., Stegun, I.A.: *Handbook of Mathematical Functions with Formulas, Graphs and Mathematical Functions*, Dover, New York (1972)
2. Bachelier, L.: *Théorie de la Spéculation. Annales Scientifiques de l'École Normale Supérieure*, **17**, 21–86 (1900)
3. Black, F., Scholes, M.S.: *The Pricing of Options and Corporate Liabilities. Journal of Political Economy*, **81**, 637–654 (1973)
4. Bodie, Z., Kane A., Marcus, A.J.: *Investments*, 6th Edition. McGraw-Hill/Irwin, Boston (2005)
5. Chriss, N.A.: *Black-Scholes and Beyond: Option Pricing Models*. McGraw-Hill, New York (1997)
6. Cootner, P.: *The Random Character of Stock Market Prices*. M.I.T. Press, Cambridge, Mass (1964)
7. Cox, J.C., Ross, S.A., Rubinstein, M.: Option Pricing: A Simplified Approach. *Journal of Financial Economics*, **7**, 229–263 (1979)
8. Curtiss, J.H.: A Note on the Theory of Moment Generating Functions, *Ann. Math. Stat.*, **13**, 430–433 (1942)
9. de Heer, R.: *The Realty Bluebook Financial Tables*. Dearborn Financial Publishing, Inc., Chicago (1995)
10. Eisenson, M.: *The Banker's Secret*. Villard Books, New York (1990)
11. Fabozzi, F.J., Mann S.V.: *The Handbook of Fixed Income Securities*, 7th Edition. McGraw-Hill, New York (2005)
12. Guillermo, G., Kou, S., Phillips, R.: Revenue Management of Callable Products, Preprint. (2004)
13. Hogg, R.V., McKean, J.W., Craig, A.T.: *Introduction to Mathematical Statistics*, 6th Edition. Prentice Hall, Upper Saddle River, New Jersey (2005)
14. Hull, J.C.: *Options, Futures and Other Derivatives*, 6th Edition. Prentice Hall, Upper Saddle River, New Jersey (2006)
15. James, J., Webber, N.: *Interest Rate Modelling*. John Wiley and Sons, New York (2000)
16. Kellison, S.G.: *The Theory of Interest*, 2nd Edition. McGraw-Hill/Irwin, Homewood, Illinois (1991)
17. Lindeberg, J.W.: Eine neue Herleitung des Exponentialgesetzes in der Wahrscheinlichkeitsrechnung, *Math. Z.*, **15**, 211–225 (1922)

18. Luenberger, D.G.: *Investment Science*. Oxford University Press, New York (1997)
19. Merton, R.C.: Theory of Rational Option Pricing, *Bell Journal of Economic and Management Science*, **4**, 141–183 (Spring 1973)
20. Mood, A.M., Graybill, F.A., Boes, D.C.: *Introduction to the Theory of Statistics*, 3rd Edition. McGraw-Hill, New York (1973)
21. Ostrom, D.: Japanese Interest Rates Enter Negative Territory. *Japanese Economics Institute Weekly Review*, **43**, 4–5 (1998)
22. Rodriguez, R.J.: Bond Duration, Maturity, And Concavity: A Closer Second Look. To be presented at the FMA Annual Meeting, Salt Lake City, Utah, in October 2006, as reported on the web page http://www.fma.org/SLC/Papers/DurationConcavity0106FMA.pdf.
23. Ross, S.M.: *An Elementary Introduction to Mathematical Finance: Options and Other Topics*, 2nd Edition. Cambridge University Press, New York (2002)
24. Thorp, E.O., Kassouf, S.M.: *Beat the Market, A Scientific Stock Market System*. Random House, New York (1967)

Index